THE HIDDEN LIFE OF LIFE

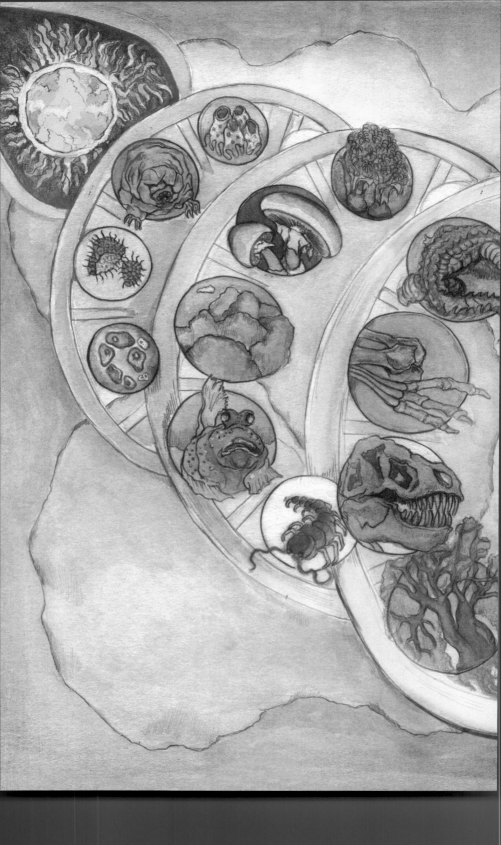

ANIMALIBUS VOL. 13
OF ANIMALS AND CULTURES

Nigel Rothfels, General Editor

ADVISORY BOARD:

Steve Baker (University of Central Lancashire)

Susan McHugh (University of New England)

Garry Marvin (Roehampton University)

Kari Weil (Wesleyan University)

Books in the Animalibus series share a fascination with the status and the role of animals in human life. Crossing the humanities and the social sciences to include work in history, anthropology, social and cultural geography, environmental studies, and literary and art criticism, these books ask what thinking about non-human animals can teach us about human cultures, about what it means to be human, and about how that meaning might shift across times and places.

OTHER TITLES IN THE SERIES:

Rachel Poliquin, *The Breathless Zoo: Taxidermy and the Cultures of Longing*

Joan B. Landes, Paula Young Lee, and Paul Youngquist, eds., *Gorgeous Beasts: Animal Bodies in Historical Perspective*

Liv Emma Thorsen, Karen A. Rader, and Adam Dodd, eds., *Animals on Display: The Creaturely in Museums, Zoos, and Natural History*

Ann-Janine Morey, *Picturing Dogs, Seeing Ourselves: Vintage American Photographs*

Mary Sanders Pollock, *Storytelling Apes: Primatology Narratives Past and Future*

Ingrid H. Tague, *Animal Companions: Pets and Social Change in Eighteenth-Century Britain*

Dick Blau and Nigel Rothfels, *Elephant House*

Marcus Baynes-Rock, *Among the Bone Eaters: Encounters with Hyenas in Harar*

Monica Mattfeld, *Becoming Centaur: Eighteenth-Century Masculinity and English Horsemanship*

Heather Swan, *Where Honeybees Thrive: Stories from the Field*

Karen Raber and Monica Mattfeld, eds., *Performing Animals: History, Agency, Theater*

J. Keri Cronin, *Art for Animals: Visual Culture and Animal Advocacy, 1870–1914*

The Hidden Life of Life

A WALK THROUGH THE REACHES OF TIME

Elizabeth Marshall Thomas

The Pennsylvania State University Press
University Park, Pennsylvania

Library of Congress Cataloging-in-Publication Data

Names: Thomas, Elizabeth Marshall, 1931– author.
Title: The hidden life of life : a walk through the reaches of time / Elizabeth Marshall Thomas.
Other titles: Animalibus.
Description: University Park, Pennsylvania : The Pennsylvania State University Press, [2018] | Series: Animalibus: of animals and cultures | Includes bibliographical references and index.
Summary: "Provides an overview of the commonality of life on Earth. Inspired by the idea of symbiosis in evolution, the book explores the challenges and behaviors shared by creatures from bacteria to humans and all those in between"—Provided by publisher.
Identifiers: LCCN 2017058555 | ISBN 9780271081014 (cloth : alk. paper)
Subjects: LCSH: Life (Biology) | Evolution (Biology)
Classification: LCC QH501.T46 2018 | DDC 570—dc23
LC record available at https://lccn.loc.gov/2017058555

Copyright © 2018 Elizabeth M. Thomas Revocable Trust
All rights reserved
Printed in the United States of America
Published by The Pennsylvania State University Press,
University Park, PA 16802-1003

The Pennsylvania State University Press is a member of the Association of University Presses.

It is the policy of The Pennsylvania State University Press to use acid-free paper. Publications on uncoated stock satisfy the minimum requirements of American National Standard for Information Sciences—Permanence of Paper for Printed Library Material, ANSI Z39.48-1992.

Frontispiece © 2017 stef lenk | www.fahrenheit450.org.

Contents

This book is dedicated to

ANNA MARTIN, NANCY FOLSOM, *and* **JANICE FROST**

1. About T...
2. Our Be...
3. Micr...
4. Pr...
5. F...
6.

16. Crocodiles / 126

17. Birds / 130

18. Mammals / 137

19. From Monkeys to the Missing Link / 145

20. The Line to *Homo Sapiens* / 150

21. Neandertals / 159

22. Why Do We Look the Way We Look? / 164

23. The San, Formerly Known as Bushmen / 167

24. Gaia's Rule One: Find a Source of Energy / 178

25. Gaia's Rule Two: Protect Yourself / 184

26. Gaia's Rule Three: Reproduce / 190

27. The Present / 194

28. The Future / 199

CODA / 201

ACKNOWLEDGMENTS / 203

Our fascination with the possibility of life on other worlds is entirely understandable. Yet it is passing strange that we pay so little attention to the largely unknown world all around us. The strange yet lovely biosphere is our only harbor in the vastness of space.

E. O. WILSON

We and our primate relatives are not special, just recent.

LYNN MARGULIS

We never think when we see a deer in a field or a bird in the sky that it's an individual with a persona—we see it as part of a beautiful view, that we're looking at "nature." And so we are, but there's more to it.

SY MONTGOMERY

CHAPTER 1

About This Book

Anthropomorphism is deeply rooted in our culture. The word comes from Ancient Greek—*anthrōp* meaning "human" and *morphos* meaning "having form"—hence to anthropomorphize is to present a non-human as if it had human characteristics. During the Middle Ages, the people in Europe believed that animals so closely resembled humans that in 1475 a sow and her piglets were tried in court for murdering a child. The piglets were acquitted, but the sow was found guilty and executed. That poor pig was on her own, but in other such trials, some animals accused of crimes were provided with lawyers to defend them.

As the Middle Ages faded, modern science bloomed. Intangibles such as memory, thoughts, and consciousness can't be measured, and by scientific standards, no hard evidence existed that other life-forms had those gifts. Lacking evidence, scientists took the position that they didn't, and the word "anthropomorphism" became a term for unscientific twaddle or sentimental folly.

The surprising thing, at least to me, was the extremes to which this notion expanded. I once read a paper about a male leopard, written with strangely twisted sentences so the author could avoid calling him "he." If you study a leopard's accomplishments as this author did, you'd find it painful to call him "it" as if he was a "thing," but calling him "he" was out of the question because "he" was used only for people. The poor author wanted to publish her paper in an accredited scientific journal, which was where I saw it, and thought it better to write unreadable prose than to disgrace herself with anthropomorphism.

But times are changing. Today, although we may never acknowledge this, we're beginning to see that the people in the Middle Ages had it right. A contemporary scientist, the biologist and primatologist Frans de Waal, pushed back hard against viewing non-humans as "things" by naming the concept "anthropodenial"—viewing a life-form as if it did *not* have human characteristics. We won't be calling our enlightenment "anthropomorphism" because we'll probably call it "science," and we won't be putting pigs on trial or finding lawyers for them—we'll just shoot them—but we're beginning to acknowledge that we share an astonishing number of characteristics not only with other animals but also with fungi and plants.

We're a recent, tag-end species, brought here by steps taken by our ancestors and reaching back through time. We share characteristics with some of the earliest life, to say nothing of those who followed. It's true that our form diverges from the others—all species diverge or they wouldn't be distinct—so here we're no exception; we clumped non-humans together and set ourselves apart because we didn't seem to credit where we came from.

To know a crow from a raven or a spruce from a hemlock doesn't mean we understand the qualities we share or how and why we came by these qualities. We also don't appreciate the marvels achieved by other species, even the most familiar, such as houseflies or bumblebees, with complex abilities and evolutionary pathways that would confound us if we knew about them. We're all doing the same thing, trying to live as best we can among the multitude of life-forms that covered our planet for much of its existence. We're deeply interrelated, we share a common ancestor, we're tied to one another, and here I try to trace the way this came about.

A reader who isn't a scientist will surely be familiar with some of the information in this book and when coming upon a familiar fact will say, "Pshaw, I knew that already." But many may find some material new and perhaps surprising, as this author was to learn after questioning her best-educated, most sophisticated, non-scientist friends. None of them knew what a "eukaryote" was, for instance, even though these friends are themselves eukaryotes and will be so for life. In fact, their eukaryote condition defines them. It's by far our most important feature but is generally considered to be beyond a layman's grasp.

This book is written by a eukaryote with no special training and is aimed at people like herself and her abovementioned friends. No doubt we were passing notes or otherwise diverting ourselves in high school science classes and now don't burden ourselves with information we think we'll never need. And even if we want such information, it is normally presented in scientific language and understood by relatively few.

I mean no disrespect for science. If not for science, a book such as this would consist of one blank page. But more people understand Yoruba and Ngakaramojong than understand science-speak. For instance, "The sporangia are typically globose to reinform, with dehiscence generally occurring along the distal edge."* Did this tell you something new about a plant? If you try to learn about a dinosaur, you're told that it's an archosaur, and if you try to learn what that is, you're told it's a diapsid amniote. Whatever that means is private for scientists, and you are not included.

But this need not be so. Scientists study and explain the various life-forms, but they don't own them. The life-forms of the past three billion years belong to all of us. In them we find our ancestors and the

---

* Chosen at random from Thomas N. Tucker, *Paleobotany: An Introduction to Fossil Plant Biology* (New York: McGraw-Hill, 1981), 104.

relatives of our ancestors. The smart ones gave rise to the scientists, but in the end, we're all parts of one thing. So I aim for a layman's vocabulary. For instance, while nothing can be done about the word "eukaryote," which lacks a translation other than "us," the name of the other two kinds of life-forms—the "prokaryotes"—can be translated as "microbes." Today that word is often scorned, but everyone knows what it means.

Many scientific names are unpronounceable. I try to offer the pronunciations in parentheses, although sometimes a word or name has another pronunciation I may not know. I also add the meaning after the pronunciation when I can—eukaryote (you-CARRY-oat, "good kernel"), for instance. But this can't always be done. Not all translations seem accurate, and not all names come from Latin or Greek. And many names have more than one translation.

I rarely call a life-form "it," since I don't share the fears of the scientist who wrote about the leopard. Here, if a life-form has gender, the pronoun is "he," "she," or "who," never "it" or "which." In the same spirit, I anthropomorphize without shame. Sometimes I call the natural world or the evolutionary process "Gaia." Or I might say a plant "remembers" something or "communicates" something, because plants do remember and communicate in ways that are different from ours but with similar results and for the same reasons. Knowledge is different from instinct; want is different from need. So if I say a plant or an animal "knows" or "wants," I mean it. Here I must confess that when avoiding anthropodenial, I may sometimes exceed what de Waal had in mind. At other times, too, I may express myself in ways most scientists will find vague or oversimplified. If this disturbs a reader, it should read a different book.

Our objections arise from our sense of importance. We've said we were made in the image of God. We see ourselves on the topmost rung of the evolutionary ladder. Of course we do. We invented the ladder. What else would we put on the top? Up on our ladder, we ignore the fact that every life-form, from a spider to a giant sequoia, is at the top of its own ladder with abilities we can barely imagine. We can't survive a massive dose of radiation or live a lifetime under water. We can't inhabit a glacier or another living organism. We can't spin a web or make the silk to do it (although we've tried) even though it's acknowledged as the world's strongest natural fiber. We can't see in the dark, we can't hibernate, we can't kill a giraffe with our mouth or hold our breath for half an hour, and we don't know how to split ourselves in half. We can't even metabolize a biochemical with sunlight, although both are readily available. And once you learn that this implies using parts of the sunlight to take atoms from molecules, you wonder why, when naming ourselves, we followed *Homo* with *sapiens*, which in English means "wise."

Lynn Margulis, the late, world-famous scientist, addressed our self-regard as follows: "There are no 'higher beings,' no 'lower animals,' no angels, and no gods. Even the 'higher primates' are not higher. We and our primate relatives are not special, just recent. We are newcomers on the evolutionary stage."*

A discussion of the evolutionary stage involves theory, which has several meanings ranging from "just a guess" to "set in cement," so certain theories mentioned here may be modified while others may stand fast. Evolution, for instance, is a modified theory, once held to be a flash event that took place in about a week, now theorized as a multi-billion-year process that brought us here and is ongoing. Those who call this "the theory of evolution" mean one thing, and those who say "it's just a theory" mean something else.

Theories are modified or cancelled, new ones arrive, and those mentioned in this book may vary in probability. Wherever I can, I try to signal this. Science is as important to the world as water, and everything resulting from its liquid discipline has a "maybe" tucked in somewhere. That's the great thing about science. If not for the "maybe," science would freeze.

---

* Lynn Margulis, *Symbiotic Planet: A New Look at Evolution* (New York: Basic Books, 1998), 3.

Then there's perception of time. To grasp the essence of the evolutionary stage, we must imagine the years of its existence in millions and billions, which isn't easy. I noticed the problem while trying to imagine the edge of the universe, said to be 13.8 billion light-years away.

As a child, I was told that heaven was beyond the universe, and that about two thousand years ago, Jesus left the earth to go there. At the time, I thought he must be looking down at us from heaven, but now it would seem that he's still on his way. Even if he's going at the speed of light, or 186,000 miles per second, at the time of this writing he's only gone as far as the Snow Angel Nebula so he's still in the Milky Way. That's nowhere near the edge of the universe. His journey has just started.

The image isn't meant to fit with religion—heaven could be somewhere else—and it certainly doesn't fit with science because the universe is expanding, and who knows what that would mean to a space-traveler trying to reach the edge? It simply suggests how dim our perceptions can be when involved with long distance and time.

My own perceptions of space and time were improved by my wonderful cousin, Tom Bryant, an astronomer who specializes in double stars. He's a teacher at heart, so when he visits, he brings a big telescope to familiarize us as best he can with what's in the night sky.

He set up the telescope in our driveway and with it showed us Jupiter with its string of moons and strips of storms. Then he showed us the moon with its craters and mountains as yet undamaged by our species. He showed us the Andromeda galaxy, which appeared as a tiny fuzzball in the eastern sky but is really a collection of a trillion stars—yes, a trillion stars—with a black hole among them, and in the northern sky he showed us Camelopardalis, a constellation that circles the pole. In it was a tiny, pinpoint dot that was a quasar, four billion light-years away. It looked like a star but was really an enormous, mysterious ball of energy, perhaps with a black hole inside it.

Tom showed us another quasar in the constellation Draco that was 8.5 billion light-years away. But the one in Camelopardalis seemed more compelling, because the light that came in our eyes that night was starting toward us when life began on earth.

That's a long time. My skin prickled. Tom showed us a galaxy in Virgo. The light we saw from that galaxy left sixty-six million years ago, he said, which means it left at the time of the Cretaceous-Tertiary or K-T extinction.

Again, my skin prickled. That extinction ended the reign of the dinosaurs, one of whom was a certain *Tyrannosaurus rex*, whose fossil I saw in a museum, an assemblage of giant bones arranged in pre-attack position and topped by a horrifying skull with big, dagger-like teeth. The speck of light we saw with the telescope could have left its galaxy when this *T. rex* was alive. Her fossil was found in Canada, not too far from where I live now, and maybe she saw the space rock, nine miles wide, as it sped toward the Gulf of Mexico, where it crashed and caused the extinction.

But even if she just saw the night sky, that galaxy was in it, and the light it was sending when she saw it was arriving in Tom's telescope as we watched. It traveled for sixty-six million years before it reached us that night, but somehow it did connect us. What keeps us apart is time.

This has nothing to do with astrology, as a famous astronomer pointed out when asked what his sign was. If I were asked what my sign was, I'd say "I'm a Virgo." He said "I'm a Feces." That's science-speak for "shit" and it sounds like Pisces, a constellation favored by astrologists.

I think he meant that stars don't make us, and I'm sure he's right. When the extinction occurred, we were rat-sized animals who lived in the trees, and if starlight determined who we are, we'd be rat-sized today because that starlight is just now arriving. Stars don't make us. But they do connect us, commemorating all who ever lived on earth, and starlight lets us feel this.

Don't bother wishing on them though. You say "Starlight, star bright, first star I see tonight, I wish I may, I wish I might, have each wish I wish tonight," and then you make your wish, which can't possibly come true because the first "star" you see on any night is a planet. It looks around to see what star you're talking to, sees none, gives a shrug, and lets your wish fly past. Your little wish could travel for thousands of years before it meets a real star and by then it's too late. Better to wish on the sun.

CHAPTER 2

Our Bedroom

My husband and I are getting on in years, so we go to bed early. During the night, I need a drink of water. While running the faucet, I notice something dark by the shower, down near the floor. A little mildew is growing. Telling myself to fix it in the morning, I drink my glass of water and go back to the bedroom, where my husband is asleep. Snow has fallen, reflecting the moon and filling the room with pale light. Our dog sleeps between us, and she woke when I got up, so she smiles and pats her tail twice to greet my return. One of our cats sleeps on my pillow because our bed is under the window, and he likes to know what's happening outdoors. He's also awake, and he purrs when he sees me. Then a new thought strikes him and he jumps to the windowsill. We live in rural New Hampshire, and he may be wondering about a coyote who goes by now and then. If the cat is interested, so am I, and I also look out the window. No one is there that I can see—just moonlight and fresh snow.

I get in bed and the cat comes with me, delicately side-stepping a small potted plant on the windowsill. He brushes against it but doesn't knock it over. For his care, I'm grateful. The plant is a hyacinth, the gift of a friend. A fall to the floor could damage it, and that would traumatize me because I've been caring for the bulb since it sprouted. It will soon start to bloom, so I'm happy it's safe, all in one piece with the soil still packed tightly around it.

I pull up the blanket and the cat curls in beside me. He puts his hand on my arm and rests his head against me. His purring fades to soft, intermittent little purrs. He's dozing off. Soon all of us are sleeping, quietly breathing, and our room is a peaceful, moonlit scene with no surprises.

Who would guess that in our bedroom and bathroom are representatives of every major group of life-forms that ever lived on earth since life began?

We have a limited view of what's around us. During the day, our room is bright and sunny. Standing near the window is like standing near a heater. We know the warmth is from the sun, but it seldom strikes us that this implies the same kind of flames you get by lighting a match, except that these flames are ninety-three million miles away. What fuel is burning, and where does it come from? Did the sun begin with a full fuel ration? And if it did, how much is left? Our poor little furnace, toiling away in the basement, seems helpless and small.

And when the snow was falling, we thought of it as new, but we were mistaken. That snow was maybe ten billion years old, much older than the earth on which it was falling. And according to a widely accepted theory, the moon is younger than the snow, having been torn from the earth by an asteroid about the size of Mars.

The water that now is snow must have been on earth as liquid or steam when this happened. But Gaia didn't make that water; she just collected it. It's thought that the Big Bang made the hydrogen part of the H_2O, and maybe three billion years later the explosion of a supernova made the oxygen part. Together they made molecules of water, and according to at least one theory, these were in the space-dust of the supernova as its swirling particles combined to form our solar system. All our planet can do about the water now is to turn it into ice or mist, or the snow reflecting moonlight through the window.

Yet we don't think of its age when we see it, and we don't understand the range of life-forms revealed by the moonlight it reflects, or those too small to see, yet here we all are, all in one room, someone from every major kind of life-form that ever graced our planet, me with a ten-billion-year-old substance in my stomach, and the issues of a civet-like Eocene mammal from sixty million years ago asleep on either side of me. The issue who is now a dog sits up and turns toward the window, her ears stiff and her eyes wide. Perhaps the coyote, another issue of that civet, just went by.

What are these groups of life-forms, and why are we in them? Uppermost are three Domains: one for bacteria, another for archaea* (which are microbes that look and act like bacteria but are different inside), and the third for us, the eukaryotes.

Our Eukaryote Domain has four Kingdoms—one for animals, one for fungi, one for plants, and one for protists, the catch-all name for a life-form that isn't a microbe or one of the above.** Amoebas are protists, for example, and those in our room are with the hyacinth. The soil in its pot is wild from our field where protists abound, not pre-treated soil from a store. The animals in our room are the dog and the cat, me and my husband, a few little hunting spiders, a few little web-spinning spiders, and probably some dust mites. The plant, of course, is the hyacinth, and the fungus is the mildew I saw in the bathroom. So maybe I won't clean it up after all. It has taken shelter in the house where I take shelter. It is clinging to life, just like me. Such different lives! Such different organisms! In the morning, one will feed itself with sunlight. Another will write about ethnic conflict in Czechoslovakia after World War I. A third can see us even in the dark, because whatever glimmer comes into his eyes bounces through them again so he gets twice as

---

* ARR-kee-ah or arr-KEE-ah—"ancient beings."
** The name "protist" is falling from general use because so many different kinds of organisms are included in that category. The Protist Kingdom is being divided, and its members are being assigned to different groups, but I continue with the name because I'm a non-scientist who needs it.

much light. Others are preying upon the dust mites who prey upon skin particles the rest of us have shed, and still others find nourishment in drywall.

But those are just the larger life-forms. Most of the organisms in our room are invisible but nevertheless are everywhere, busily splitting themselves in half. These, of course, are microbes. Some are in our stomachs, breaking down the food we ate for dinner so we can absorb its nourishment, and in return for their help we're keeping them fed and wet.

Thus our relationship with them is symbiotic. Two organisms have joined for the benefit of both. That we have this relationship is not at all surprising—almost everything living on the planet has this relationship with something else, because symbiosis has been practiced since the earliest times. Cooperation is an important engine of survival, and all who manage to do it are better off than they would be if they were going it alone.

Helpful microbes are not our only occupants. Inside our eukaryote bodies, our own cells are outnumbered by microbes ten to one. We don't know what they are or what they're doing, and we never think about them. We see ourselves as independent solitary organisms when in fact we're heavily populated ecosystems, filled to the brim with passengers as if we were overcrowded trains.

So there you have it—representatives from all the major groups of organisms that ever existed on our planet, right there in our bedroom and bathroom. We may seem a bit different in the ways we look, breathe, move, and metabolize, but we're related. Our genes may be arranged in different sequences but we share a common ancestor, so we pass genetic information with the same genetic code. Or almost the same, as a somewhat alternate theory has recently been offered. This being the case, what's so surprising is how little we know about the others who evolved from our ancestor. And that would be a microbe who lived in water, maybe in the sea.

CHAPTER 3

Microbes

When light we see today was leaving the galaxy in Draco, Gaia was planting the tree of life, which would grow to have branches made of ancestors and their descendants. And to prevent her creations from going to waste, she made three rules for all. (1) *You must find a source of energy to keep yourself going.* (2) *You must protect yourself from the elements and from those who see you as a source of energy.* (3) *And I, Gaia, got you started, but it's up to you to make others like yourself.*

These rules went into effect four billion years ago or even earlier, and ever since then these rules have been observed—not always by everyone (by now, most of the species that ever existed—more than a billion—have gone extinct). But every life-form with us today has an unbroken chain of ancestors, one individual after another, that reaches all the way back to a single cell, the first, the earliest, life.

Even so, we seldom think when we see an oak tree or a seaweed, a mushroom or a fly, that it has a parent whose parent had a parent who in turn had a parent, going back for billions of years to a single cell, the tiniest possible common ancestor, and that hour after hour, day after day, ever since the day that life began, every one of those ancestors kept Gaia's rules. This is true of us and everything we see that's animate. We're all made entirely of cells, nothing but cells, which are much like the first. They're called "the building blocks of life." And they all descended from a single cell, somewhere in the water.*

In the report of an Origin of Life conference at Princeton University in January 2013, Dr. Steven Benner, a professor of chemistry and microbiology, is quoted as saying "When it comes to the origin of life, there are no experts." But theories abound, one being that molecules necessary to form a living organism were brought in by comets or meteors, a theory that was reinforced in 2012 when astronomers in Denmark found a star in the process of forming just as the sun and its planets had formed. The right kinds of molecules were detected in space-dust contributing to this event. Yet another hypothesis known as Panspermia suggests that life existed throughout the universe in the form of highly resistant organisms that were present in the space-dust at the time our planet formed.

Then too, the first living cells could have formed from earthly molecules, which seems to be the mainstream theory now. It holds that life started in the ocean, although more recently a fresh water theory arose. Molecules and atoms combine as they can, and early in the earth's history such molecules may have formed a little string that caught the attention of Gaia.

Whatever it was, it wasn't a cell and was tiny and alone, perhaps in the vastness of a world-wide, perilous ocean. It must have had a source of energy, perhaps from sunlight, perhaps from an undersea vent, so it must have chanced upon Gaia's first rule. But almost anything could have torn it apart unless it had protection—something around it to keep it together and keep other molecules from interacting with it.

---

* The smallest cell now known is a parasitic bacterium, *Mycoplasma galliceptium*, who lives in primates such as ourselves, and is roughly a millionth of an inch in diameter. The largest cell is an ostrich egg before it's fertilized—about six inches long, five inches wide, and weighs about three pounds. It's the same size after it's fertilized, but by then a sperm cell has joined it so it's two cells.

You're likely to vanish if you don't keep Gaia's rule for self-protection. Somehow the little life-form found a cell wall.

One theory holds that the cell wall came first, perhaps as a bubble in which RNA had formed or gathered. That, along with DNA, is a kind of acid that transmits genetic information, but how acids are able to do this is almost beyond comprehension. You need a clear knowledge of chemistry, biology, and maybe even physics just to read about them, yet these made the first forms of life.

Certain kinds of clay can also make bubbles that self-replicate, so the clay seems to grow as if it were alive. No firm theory addresses this question, but perhaps whatever components made RNA were in some of that clay or in something like it when it bubbled. On the other hand, perhaps not. A favored theory holds that the cell wall came long after any string of molecules appeared that could be called "life."

At any rate, the cell wall made all the difference. A strand of RNA just floating around could almost describe a virus, which is little more than a string of unprotected molecules and therefore can't keep Gaia's third rule, or not as we understand it, and make another virus on its own. Because of this, it's said not to be a life-form. Even so, it might look like the earliest life-form and possibly act like it too.

True, it doesn't reproduce the way other life-forms do. Instead of splitting in half or inserting its genetic material into someone else, it enters a cell that will act on its behalf—one in your mouth, maybe—and uses your cell's process for itself. This doesn't mean the first life-form did anything similar, but it does show that even a string of molecules can do what's needed to survive.

Or so it seems to me, a non-scientist who gets amazed by things. I look out the window and see grass, bushes, two crows in a tree, and a deer in my field. The grass doesn't seem much like the deer who is eating it, but if you could look back in time, viewing their ancestors, you'd find increasingly fewer ancestors who were more and more alike, until you came to a little string of molecules that looked like a virus. So perhaps the closest we can ever come to knowing what the little string was like is by knowing a virus.

Although we seriously disrespect them, we've all had encounters with them, probably like one of my own when I was standing in the kitchen and saw a crumpled tissue on the floor. I picked it up. It looked fairly clean and nobody was sick, so how was I to know it was infected? After throwing it away I must have rubbed my eye, and on my finger was a virus, smaller than a microbe, but even so, it's the hero of my story. To the virus, the area between my eyeball and my lower eyelid must have seemed like the Grand Canyon full of water would seem to one of us, but the virus managed to drift through it and slip into one of my cells.

Again, the size difference must have been considerable—if the cell in my eyelid was a hot air balloon with a motor and passengers, the virus would have been a tennis ball or maybe even a grape, so the cell didn't know it had internalized the virus, perhaps not even after the virus spread its genes throughout the cell, which then released a cloud of viruses as if they were of its own making. This is why viruses are said to be more closely related to their hosts than to each other.

If an eyelid cell births a cloud of viruses, it does seem like their mother, but I'm not sure I like being related to a virus, especially not the one who made me sick, or that a cell in one of my own eyelids would allow this. What was it thinking? If I get sick and die, its fate is sealed along with mine. Soon my body was swarming with viruses.

What happened next must have been amazing, although I knew nothing about it, because the "I" with whom I'm familiar is just who I see in the mirror with a collection of memories and thoughts. My brain knew, though. It sensed something.

Most of my brain has nothing to do with me or my thoughts because it's preoccupied with my body. Unknown to me, when it noticed all the viruses, one of its departments began to call upon the white blood cells who had been waiting inside me, even in my bones.

They began chasing around, hunting the viruses, and soon so much was happening, what with chemical signals, temperature signals, white cells, muscle cells, excitable neurons and all, that an encyclopedia devoted just to this phenomenon would be needed to describe them. That was me?

The me with whom I'm familiar was slumped in a chair in the kitchen, wondering why she didn't feel so good. Soon enough I had a fever. My body was trying to burn up the viruses, doing its best to save itself and me as its passenger with it. Then, somewhere in my brain, something must have noted that more help was needed, and a massive transfer of information must have begun between my brain and my chest because my chest muscles squeezed and I coughed.

By then I'd gone to bed because I felt so awful, so I didn't completely cover my mouth. The vapor of my breath was carrying viruses, which I unwittingly threw into the air. They drifted over surfaces like the blanket and the bedside table to wait for someone else. It's my understanding that a virus can sit on a table all day and still be vital, if that's what you call it, so soon enough, with any luck, a host would pick it up.

It's amazing to think that the cough (I'm not speaking of my body, just my cough) resulted after billions of years of evolution from something like a virus, advancing step by step with lengthy trials and errors to become an organism like myself with different parts and functions and masses of internal systems that I, their owner, don't control or understand. But that's evolution for you. Something as complicated as a cough began with a tiny group of molecules, not unlike a virus, that somehow strung themselves together, probably in the sea!

If only the little string had left a fossil. It didn't, or none that's been found. But other single cells left fossils, unlikely as that may seem. Whether on land or in water, a living cell gets covered with sediment and disintegrates, leaving only the minerals it might have contained or those that leaked into it somehow. After a very long time, usually under heavy pressure as more sediment is deposited above it, the layer with the cell solidifies until the cell itself, however small, is something like a rock.

The earliest fossil known to have had life began as a microbe who became a fossil 3.7 billion years ago. Several million years later, a group of microbes had evolved to the degree that they gathered together in a mat that became a fossil in what is now Australia.

Such microbes invented photosynthesis, making their nourishment with energy from the sun or from an undersea vent where fire from the earth's core comes up through the bottom of the sea. They found ways to take apart molecules of air or water—probably at first it was water—looking for carbon, hydrogen, and oxygen, with which they make an elegant sugar, $C_{12}H_{22}O_{11}$, which is sucrose, their food. But whether they get their hydrogen from water (H_2O) or their carbon dioxide from the air (CO_2), they don't use all the oxygen (O), so they toss what's left. We found a use for their waste, so we developed ourselves in such a way that we can get it by breathing.

Thus microbes invented one of the most important engines of evolution that eventually covered the world with millions of different life-forms. We think of ourselves as the inventors. We point to fire, the wheel, the H-bomb, and the iPhone. But if those microbes invented photosynthesis, can our efforts compare with what they did before they became a fossil?

The earliest life-forms with us now may be the microbes known as archaea, the "ancient beings" mentioned earlier. Once these were thought to be bacteria because that's what they look like—taking all kinds of shapes such as rods, spirals, capsules, and plates depending on the species, none with a nucleus, all with their genetic materials and biochemicals just floating around inside them, until it was discovered that the biochemicals and processes inside archaea were quite different from those inside bacteria and, interestingly enough, were more like ours.

We didn't develop in the same direction, though—archaea do things we can't imagine doing, and can live almost anywhere including the most unlikely places: hot springs, glaciers, salt lakes, acids, rocks, other life-forms, and oceanic vents. In other words, they're hardy.

By now there are thousands of kinds of archaea. Some have a tiny tail that whips back and forth, propelling its owner through whatever moisture it inhabits. Others have tiny, hair-like features on their cell walls, which they use to pull themselves along. An important theory holds that these resulted from symbiotic relationships between two kinds of archaea, whereby tiny, thin archaea joined themselves to somewhat larger archaea to become little tails or hairs. All parties must have benefited, and the newly created life-forms seem to have flourished.

Something like that may be seen in us. The mitochondria in our cells are thought to have been independent organisms once—they're different from our body cells in that their genes are like those of bacteria rather than like ours. Now in our cells, they function as little generators, producing chemical energy. One imagines a mitochondrion moving into a eukaryote cell, perhaps in search of shelter. The eukaryote cell tries at first to evict it, but then experiences some pleasant change—a little burst of energy, perhaps—which seems so helpful that the eukaryote decides to leave the mitochondrion alone. No doubt this description is overly anthropomorphic, which here I've carried to extremes, but when it comes to life-forms, we're an adequate example. Anthropomorphism may not be perfect, but anthropodenial is worse.

According to several theories, millions of years went by before bacteria appeared, but it's fairly well accepted that at least 2.7 billion years ago, they were present. They have the same basic genetic code as archaea, so they might have started in the same way, or they might have descended from archaea, assuming, perhaps wrongly, that archaea came first. At any rate, although bacteria seem to have their own evolutionary history, they look like archaea—single cells with a cell wall but no nucleus—and they act like archaea, although they're not as tough and are seldom found in the extreme environments where archaea often thrive.

Like archaea, some bacteria have tiny, hair-like features for attaching themselves to objects, including other bacteria. And again like archaea, some bacteria gather together as films or slimes when conditions are unfavorable. Sometimes one kind of bacteria joins the slime, and sometimes many kinds join it.

Some bacteria advanced beyond archaea, and these may have changed the world. Mycobacteria,* for example, band together to form slimes called "wolf-packs."

---

* *MIKE-o-bacteria*—"slime bacteria."

These wolf-packs don't just wait in one place for moisture or nourishment to come their way. They slide to different places in search of what they need. If conditions continue to deteriorate, the wolf-pack may bunch up to form an extremely small body that looks vaguely like a mushroom although barely visible to the naked eye, in which the occupants have a division of labor. Some sacrifice themselves to become the husk or protective outer surface, while others become spores.

Often enough, the term "spore" refers to a spore such as that of a fungus—something like a seed but not a seed that holds the fungal offspring. Bacterial spores are the bacteria themselves who, when they don't like what's going on around them, scrunch up inside their toughened cell walls to wait in a dormant but stable condition until their environment improves, at which time the spore reverses and becomes a functional bacterium again.

As for the little wolf-pack of bacteria, when the time is right, their tiny structure tips over and the spores come out. They resume their bacterial forms and go their own ways, splitting in half as they do so.

For a microbe, that's compelling, but that's not all. Even though a wolf-pack disintegrates later, for a while it behaves like a more complicated, multi-cellular life-form. Here again, bacteria were pioneering a process. Multi-cellularity progressed step by step, but it's interesting to think of single-cell bacteria forming a mass that resembles a multi-cell organism, then forming spores to be released as if they were the mature organism's offspring. Today the world is filled with multi-cellular life-forms that produce multi-cellular offsprings. Did this bacterial ability produce the one-piece organism made up of many cells?

Perhaps no one knows, but here's the take-home. When Gaia sees that something works, she promotes it, not always in our direction, of course, because she cares about all her subjects. But she has always liked microbes, and perhaps while wondering if more could be done with the successful-spore concept, she encouraged some bacteria to make endospores, a kind of spore that can survive a long time without nourishment. That's Gaia's rule for self-protection taken to its furthest extreme.

Today, the oldest living thing in the world is not a bristlecone pine or a Galapagos tortoise—it's a bacterium who is spending time as an endospore. The journal *Science* published an account of *Firmicutes*[*] endospores found in the Dominican Republic inside an extinct species of bees, which, perhaps forty million years ago, were preserved in amber.[**] How did this happen? The bees were trapped in sticky resin oozing from a tree. They couldn't escape, and the resin smothered them. When their struggles proved useless and they died, the bacteria inside them sensed that things had gone wrong and spent the next few hours turning into endospores. Their plan was to wait for things to improve.

Where were we when this happened? On the day the resin trapped the bees, we were African monkeys up in the trees, skillfully swinging from branch to branch, following our own path of evolution, never to achieve what those endospores were even then achieving, however much we might want to.

It took about a century for the resin to harden into amber. The endospores were waiting, but their wait had just begun. We weren't waiting, though—during the forty million years that followed, we morphed from monkeys to great apes, left the trees, turned into human-types, learned to walk on our hind legs, shed our fur, figured out how to make tools by sharpening stones, learned how to use a fire and later how to start a fire, which was just as well because an ice age came, making glaciers that froze most of the world's water, with the result that the world didn't get enough rain.

The forests withered and the grasslands spread. We were people by then, on the African savannahs where we lived by hunting and gathering. Eventually we walked all over the world, learned how to raise plants and animals, learned how to make buildings and then cities, and at last developed universities with graduate programs in microbiology, which benefited the scientists who found the amber with the bees. The

---

[*] *Fur-MICK-you-teez*—"strong skin."
[**] R. J. Cano and M. K. Borucki, "Revival and Identification of Bacterial Spores in 25- to 40-Million-Year-Old Dominican Amber," *Science*, May 19, 1995.

scientists removed the bees and looked inside them. They found the endospores and removed them. Four hundred thousand centuries had passed since those bees were trapped in amber, but the endospores were ready. They resumed their bacterial forms and, just as if nothing had happened, went on with their little lives.

Microbes reproduce by splitting in half. This involves no genetic transfer and results in two identical single cells where at first there was just one. However, some of kinds of archaea as well as some kinds of bacteria acquired tiny hairs called pili.* Equipped with a pilus,** the more complex microbe could cling to a surface or pull itself forward.

But bacteria do more than that. Some have special pili, maybe two or three among the other little hairs. These they can push into other bacteria, not necessarily the same species, and the two can exchange genetic material through the little tube. This is known as sexual reproduction, even though bacteria are neither male nor female, and even though the two bacteria will keep splitting in half, as will the two halves, the four halves, the eight halves, and so on to infinity. But every now and then, thanks to the little hair-like pili, they manage a genetic change, which may or may not profit them. A profitable change, as far as they're concerned, would be indifference to the antibiotic their human host is taking to stop the symptoms they're causing.

As far as the phenomenon of life is concerned, the ability to exchange genes and gain variety was almost as important as the Big Bang, at least for those who live on our planet. Without this, mutation would be the only road to variety, and mutation often does more harm than good. Mutation happens when our cells are splitting, but sometimes they do it wrong. Then the DNA replicates strangely. Mostly this makes no difference, although sometimes it can cause problems, but the occasions where it's helpful are rare. So if evolution were fueled by mutation alone, might the world still be waiting for the "higher" organisms?

* *PIE-lye*—"hairs."
** *PIE-lus*—"hair."

As for sexual reproduction, although the introductory form is found in bacteria when they stick their little pili into one another, it is seldom if ever found in archaea, or so it's said, although scientists disagree about how archaea use their pili. At any rate, despite the fact that archaea can do things that confound bacteria, such as live in acids or glaciers, the bacterial use of the pilus contributes to the concept that bacteria are more "advanced." That's our view, certainly. As far as we're concerned, flourishing inside a glacier is not an advancement, but sexual reproduction is.

The importance of microbes cannot be overstated. Lynn Margulis, being a scientist, called microbes "prokaryotes" and had this to say about them: "So significant are prokaryotes and their evolution that the fundamental division in forms of life on Earth is not between plants and animals, as is commonly assumed, but between prokaryotes and eukaryotes. In their first two billion years on earth, prokaryotes continuously transformed the Earth's surface and atmosphere. They invented all of life's essential, miniaturized chemical systems—achievements that so far humanity has not approached. This ancient high biotechnology led to the development of fermentation, photosynthesis, oxygen breathing, and the removal of nitrogen gas from the air."[*]

Their numbers are incomprehensible. When one considers that the microbes in any one person's body are said to outnumber the person's body cells by ten to one, and that one human body is said to have as many as 10^{15} body cells (that's one quadrillion), this would mean that ten quadrillion microbes could be found in just one person.

They form the largest part of the earth's biomass. Since we can't see them, that's hard to believe, but even so, everywhere you look, everything you see, is loaded with them, coating the landscape, deep

---

[*] Lynn Margulis and Dorion Sagan, *Microcosmos: Four Billion Years of Microbial Evolution from Our Microbial Ancestors* (Berkeley: University of California Press, 1986), 29. Margulis held that archaea and bacteria belong to one Domain, and that archaea were a kind of bacteria.

in the soil, up in the clouds, out in the sea and in fresh water, all over the world. Their number has been expressed by 5 with 30 zeroes after it,* or five million trillion trillion, a number known as "one nonillion," not that the name means much to most of us except that it means something big. Only atoms and molecules are more abundant than microbes.

But whether we like it or not, these micro-organisms are the foundation of life. And whether we know it or not, cellular composition is the template of life. Archaea and bacteria look and act so much alike that once they were thought to be the same. But these two kinds of life-forms were found to be so different that they occupy two of the three Domains in which life-forms are classified. The genes and metabolic pathways of archaea are more like ours than like those of bacteria, so no matter how much archaea seem like bacteria, in ways they are more like us.

In contrast, we belong to the same Domain as trees, amoebas, sponges, seaweeds, ostriches, and moss. To some, our similarity to a moss or an amoeba might seem unlikely, if only because most of us seldom think about our cell structure or the organization of our DNA, or the near-infinite, ever-changing evolutionary lines that brought us here.

Our eukaryote Domain with its millions of members is now imposed over the others—the invisible ones whose territory is the whole world and everything in it, from pole to pole, from the clouds to the bottom of the sea. And much of this resulted from cooperation—symbiotic unions of two or more partners, each successful by itself, but more successful together.

Survival of the fittest was once thought to be the key to evolution—nature red in tooth and claw was thought to be the norm. But Gaia likes cooperation. It helps with all three of her rules.

---

* That would be 5,000,000,000,000,000,000,000,000,000,000 microbes in the world.

CHAPTER 4

Protists

Our kind began with microbes and their little tubes, the pili, which they stick into one another. With these, perhaps two billion years ago, they made a new kind of cell quite different from their own. The new cell had a membrane wrapped around its DNA. That membrane and its contents changed the way the world works and gave us the name eukaryote, "good kernel."

Strangely, however, the microbes took two billion years to get this done. One wonders why. If microbes already had little hair-like pili to stick into other microbes, as well as other little hairs to pull themselves around, it would seem that with their symbiotic relationships they must have been making different life-forms for a very long time. Does this mean that for two billion years they were just making new kinds of microbes?

True, there are many kinds of microbes, so their time wasn't wasted, and perhaps during all that time they did make some kind of eukaryote or even some life-forms that were neither microbes nor eukaryotes. But the earth had a rough time at first, when volcanoes were erupting and meteors were crashing down. So if the microbes produced a different, unknown kind of organism, it vanished without leaving a fossil that's been found.

The new, surviving organisms were single cells, but these cells were different. With the "good kernel" that made them eukaryotes, they were the first of the Protist Kingdom. The amoebas mentioned earlier—those in our bedroom with the hyacinth—are protists and could be described as tiny, shapeless lumps too small to see. They're bigger than microbes, but they're very much like microbes—each is just a single cell with a nucleus. But unlike a microbe, an amoeba is covered with a flexible membrane instead of a firm cell wall. Otherwise, an amoeba has no distinguishing features such as a front or rear end, but thanks to the flexible membrane, its body can bulge out to make "false legs," also known as pseudopods.* These the amoeba uses to move around and to feed itself if it comes upon a bit of organic debris or a diatom alga. It will ooze a false leg around the food until the far end touches the rest of its body, making a pocket to enclose its prey. It will then squeeze out biochemical solvents to break down the food item and will absorb the nutrients through its membrane. If waste results from the process, the waste is pushed out the way it came in. The act of an amoeba feeding itself is known as phagocytosis** ("cell eating"), and the bit of food when inside the amoeba is known as a phagosome ("eaten lump").*** After eating a lump and carrying the meal to its conclusion, the amoeba searches for more food.

But here, one must stop and wonder. A single cell, smaller than a pin point with no brain and no sense organs as such, can somehow distinguish a tiny speck of an edible substance and can reach around it, enclose it, and digest it, and then can somehow realize that what's left is waste and push it out through its membrane. An amoeba is little more than a tiny dot of cytoplasm—that's the thick juice inside a cell but not in the nucleus—yet has abilities we may admire but do not possess.

* SOO-dough-pods.
** FAG-oh-sy-TOW-sis.
*** FAG-oh-soam.

Amoebas come in many forms. Our white blood cells are a form of amoeba, except that they live inside us and can't live on their own. They start in our bone marrow but congregate in our lymph glands, although some are always in our blood. More appear when needed, to prowl our bodies like leopards in a bushland, hunting down invaders who would do us harm. We are their ecosystems but they don't know much about us, which gives them an important human quality—we don't know much about our ecosystems either.

Then there are social amoebas such as *Dictyostelium discoidium*,[*] also known as slime mold, found in damp soil and moist leaf-litter. Although these are eukaryotes just like us, they behave in much the same way as the colonies of bacteria mentioned earlier, the so-called "wolf-packs" which form slimes, move to better places, then form into mushroom-like structures with caps in which some become spores.

D. discoidium do the same thing but have taken the practice higher. When their surroundings displease them, these amoebas give off a chemical signal, which draws groups of them together. No shapeless slime for these little protists—they gather into a ball which then stretches out to become long and thin. It is now a multi-cellular creation with a front end and a rear end, and it looks like a tiny slug.

It then moves away as a slug would move, sliding along, front end forward, leaving a little trail of slime as it searches for more food. When it comes to a favorable place, its rear end spreads out and its front end rises, much the same as the wolf-pack microbes, so it looks like a mushroom with a thin stem supporting a bulb that looks like the mushroom's cap.

Again in the manner of wolf-pack bacteria, the amoebas on the outside of the structure sacrifice themselves by being the "skin." These protect the amoebas in the bulb, and when the colony is ready, the whole thing tips over and the amoebas in the bulb come out. All are now in the single-cell condition, hopefully in a better place.

Maybe a connection exists between the wolf-pack bacteria and these capable amoebas. If not, the similarity shows something that

---

[*] *DICK-tee-oh-STEEL-ee-um dis-COY-dee-um*

can happen when different organisms find similar solutions for their problems. If it worked for them, it could work for you, Gaia tells a new life-form she's making.

Other protists manage sexual reproduction and thus gain genetic diversity. A paramecium[*] is such a protist. Any protist that moves is known as a "motile" protist, and a paramecium is a very motile protist who, although extremely small, is enormous compared to a bacterium. Even so, like an amoeba, a paramecium is just one eukaryote cell, smaller than the point of a pin, possibly visible as the tiniest dot to someone with wonderful eyesight but visible to others only through a microscope. It has no gender (hence I'm forced to use the pronoun "it") but unlike the shapeless amoeba, a paramecium has a rounded front end and a tapered rear end. It doesn't have a mouth, but on its back, its upper side, it has a groove that narrows to a mouth-like pore. It has no intestine, but in its rear end is an anal pore. Our kind is slowly appearing.

Continuing some of the microbe customs, a paramecium has tiny hairs along its sides that act like oars—beating backward if it wants to go forward, beating forward if it wants to go backward, pushing itself through the water.

I had the privilege of watching parameciums through a microscope and was impressed by one of them. I wish I knew what kind it was, although to me it was an interesting little individual, not just a member of some species. However, it may have belonged to the "white rat" species, often used in laboratories for biological research because they reproduce readily and are easy to teach. That's where I saw this one.

It looked something like a miniscule banana, but thinner, oblong, tapered at both ends, and somewhat flexible. It also had two nuclei,[**] one of which, I was told, acted something like a little brain. It didn't move at first, and then suddenly shot off straight ahead, then stopped as if it thought of something, then waited a moment, then seemed to

---

[*] *Parra-ME-shum* or *parra-ME-see-um*.
[**] Plural of nucleus.

float backward more slowly. I'd heard they are good at sensing danger, and will whip around and go the other way if they detect a menace. So as best I could, I examined the space revealed by the microscope to see what might have made this little being change its mind.

I saw nothing, but that doesn't mean much. Whatever was there, if anything, the tiny creature didn't see it either because parameciums don't have eyes. This doesn't mean it didn't know things, or maybe it means it had a bad experience with something in that direction. The world of things too small for us to see is limitless.

Since then, I've read that parameciums, perhaps the "white rat" variety, can learn and remember. A considerable amount of study has gone into this with varying interpretations and results. But evidently groups of parameciums were taught to avoid a certain kind of light by receiving a mild electric shock, and the poor little things kept away from that light thereafter. Who would have thought a single cell could learn? It seems they can and do, just as we'd learn to avoid a light that gave us an electric shock. If learning is seen in parameciums, we can estimate its value. Learning must have come from the earliest times.

Parameciums do more than learn, however. They also copulate. When used for single-cell organisms, that word could irritate the science-minded, so I use it with sincere apologies and only to avoid anthropodenial.

When parameciums do it, it's a very different process than splitting in half. When splitting, a paramecium pinches itself in at what might be called its waistline and takes several long minutes to slowly pull apart. The lower half—the part that's breaking off—wiggles a little, as if it already has a future in mind and wants to speed the process. When it's free from the upper half, it seems to spend a moment collecting itself, and then it moves away.

However, when exchanging genes, two parameciums press the grooves on their backs together and stay that way briefly while their genes pass from one to the other, taking much less time than the split-in-half process. Now that we know that parameciums make decisions, or seem to, we must wonder what prompts one to find a suitable partner (hopefully not the part of itself that just broke away), turn on its side, push its back against the partner's back, and squeeze out some genes. How does it know which miniscule things inside itself are genes and not just waste matter? And how does it remove its genes from the rest of what's inside it? Or does it just squirt in whatever and let the partner figure it out?

At any rate, when they finish they go their own ways, each with some of the other's genes, and after that, each will divide in half by the traditional method, and both halves will carry mixed genes.

The lifestyles of protists are more diverse than those in any other Kingdom. Some protists are solitary; others are social. Some live symbiotically with other organisms; others are parasites of other organisms. And some don't interact with other life-forms except to eat them. Some of the most important protists were the photosynthesizing algae, who seem to have evolved from photosynthesizing bacteria.

Interestingly, virtually every fungus, animal, or plant on the planet has a protist associated with it. One that's associated with us is the enormously complicated, hard to imagine, malaria-causing *Plasmodium falciparum*,[*] a parasite with multiple life stages and two hosts, or better yet, a host and a vector—a person and an Anopheles mosquito.

A female mosquito needs some iron before she lays her eggs, and she knows it's found in hemoglobin, which is found in human blood. The parasites want glucose and other biochemicals also found in blood, so they wait in the mosquito's salivary glands until she finds a victim.

She learns of our presence by the carbon dioxide we exhale and then finds our bodies by the heat we give off. Here she must be careful because her victim will swat her, so she can't just land on him anywhere, she must land right over a capillary. She hovers near him, sniffing, until she smells the biochemicals she knows are in his blood. Aha! She lands and jabs her stinger in the capillary, squirting out a little saliva as she does. Her saliva has a juice that anesthetizes nearby

---

[*] *Plaz-MO-dee-um fal-SIP-ar-um*—"molder of a sickle shape."

nerves and keeps the victim's blood thin, making it easy to suck, but is also full of parasites. They're squirted into her victim with her saliva.

At this point in the process, the parasites are known as sporozoites,* somewhat worm-shaped but incredibly small. They're in the form they use for travel, and once inside the victim, they slide through his blood vessels with the bloodstream until they come to his liver. Here, they're supposed to enter a liver cell, but not just any liver cell. They may visit several liver cells before they find one to their liking.

The parasites aren't there to eat, they're there to get ready to eat. So they form a ball-shaped structure known as a schizont** filled with many nuclei that produce "daughter" parasites. These soon burst forth as a swarm of individuals known as merozoites.*** These are the parasite's eating form, and they look like tiny, oblong worms. They're almost ready to penetrate the person's blood cells. But they don't want the person's immune system going after them so they wrap themselves in bits of membrane from the liver cells. Thus disguised, they invade the red blood cells, which was their goal. The person isn't sick yet and has no sense of anything wrong.

Once inside the blood cells, they feed for a while on the glucose, also some of the amino acids, and perhaps the iron as well. But they destroy the hemoglobin, which has been carrying oxygen from his lungs around his body. This may let water into the cell where the parasites continue to make new ball shapes and produce more daughter protists. Water would be bad for these daughter protists. They must rupture the now-ruined blood cell so they can move out to invade healthy blood cells. They repeat the process until, at some point, thousands of blood cells rupture at about the same time, which normally happens at night. Is this intentional? How do they know it's night? But it's then that the person feels the chills of malaria.

From here, the two-fold process continues until the person's immune system deals with it, or the person takes some anti-malaria

---

* *SPORE-oh-ZO-ites*—"spore-like creatures."
** *SKIZ-ont*—"divided being."
*** *MER-oh-ZOE-ites*—"partial simple beings."

drug, or the person dies. These events would kill the parasites, but they have foreseen that possibility and are prepared. Not all have made ball-shaped schizonts. Instead, some have continued in their traveling form and didn't go into the liver. Since they're still in the bloodstream, the next mosquito can pick them up.

In time, another mosquito arrives. She jabs the malaria-stricken victim and sucks up some of his blood, which now is filled with parasites. Once inside the new mosquito, if all goes well, they will soon be in a new victim, so they, too, must prepare. They go down from the mosquito's mouth to her gut, penetrate the lining of the gut, and there become males and females. They want to exchange some genes, because they don't know the new victim. They will venture into the unknown, where genetic variety could help them. After the males have fertilized the females and new, mixed-gene parasites result, the new ones go up to their carrier's mouth, enter her salivary glands, and wait.

While all this was happening, the mosquito was flying around, searching for that victim. Soon enough, she finds a new cloud of carbon dioxide. She flies inside it, feels warmth, heads toward it, smells blood, and lands on a capillary. She jabs, and sends out some parasites. But her salivary glands are packed with parasites. Her mouth feels full. The blood she's sucking comes slowly, and if she isn't quick, her victim may swat her. Maybe she jabbed the wrong place. Trying harder, she jabs again, sending more parasites into the victim and starting the process over.

As for me, I was surprised to learn that mosquitoes had salivary glands. But if they didn't, we wouldn't get malaria. And I realize that these parasites are the enemies of our species, killing thousands of us every year, so of course I hate and fear them. Yet I cannot curb my sense of wonder when I think of what this protist has achieved. My parents endowed me with a basic form when I was an embryo, a form like theirs, which I passed on to my children and will keep until I die. So I find it hard to imagine a species with a form suited for a mosquito's gut, another for a person's liver, another for a person's red blood cells, and finally a repeated form—the ball-shaped structure—for converting some of these forms into forms who are ready to take the next step, including the move to another victim, to say nothing of the different places they must go once they're in that victim and the array of biochemicals they must identify.

A malaria parasite knows more about human blood than most humans do, and it's hard to know what to make of it. That parasite is far from simple. Never mind that it qualifies for a PhD in biochemistry; just think of its shapes. A tadpole who becomes a frog is one individual, and one amoeba who splits in half becomes two individuals. But do either of these definitions fit a malaria parasite?

No definition seems to fit, and the human experience includes nothing like it. For us it might be like a group of biochemists getting off a plane and going to a restaurant, rushing to the bathroom where they turn into dogs dressed in aprons, then into the kitchen where they snatch food from the cooks, then into a locker where they reproduce sexually and give birth to a crowd of well-educated humans who rush to the airport, catch a plane, and start over.

This wouldn't be done by an individual parasite—or not exactly—although something from each individual transforms as the next. But we can't even imagine such a thing, let alone do it, and we tend to discredit those who can.

"So what?" we say. "Those are bugs or something. All they do is make us sick. We humans are a higher form than that." In this we're like children in a playpen, very handy with our toys but unaware of what was needed to put astronauts on the moon. Making a skyscraper or a nuclear reactor is certainly an achievement, but does it compare to being much smaller than a blood cell, inhabiting different systems in two kinds of animals while achieving a predetermined series of complicated goals by taking different forms? As far as the parasites are concerned, a human is just a place to get this done, and since they're too small to crawl from one person to another, they ride in mosquitoes as we ride in planes.

Who knows how these remarkable abilities began in these protists? And how long did it take a humble protist to achieve so much?

Plasmodium is the name of their kind and includes maybe two hundred species, all with similar life-cycles and all with two hosts—one of them a mosquito and the other a vertebrate, usually a mammal. And to think that these protists—eukaryotes just like us—are so tiny that perhaps thirty of them can inhabit a single red blood cell. Does this say something about size range when it comes to our Eukaryote Domain in contrast to archaea and bacteria—the Prokaryote Domains? Not one of the microbes is big enough to see without a microscope, but as for us eukaryotes, thirty malaria parasites can fit inside one red blood cell, while a blue whale is a hundred feet long and has a tongue that weighs as much as an elephant and a heart that weighs as much as a car.[*] The prokaryotes may have invented "all of life's essential, miniaturized chemical systems," as Lynn Margulis explains, so they certainly got us going, but when it comes to development, we've got the size range nailed. And that's just for animals. Fungi and plants do as well or even better. The smallest fungus is a yeast—two micrometers in diameter—and the biggest is a giant who occupies four square miles and will be mentioned later. The smallest plant would be a duckweed—a tiny dot floating in the water—and the biggest, or at least the most massive, would be a certain giant sequoia, but at 275 feet this tree is not the tallest. A certain redwood is twenty-five feet taller.

In contrast to the malaria protists, we should consider another protist known as giant kelp, one of the largest life-forms on the planet. Kelp are brown algae and are successful in a simple way. A complex lifestyle is not for them. No, a newly formed kelp will fasten to an undersea support where it will stay for the rest of its life, reaching for the sun by growing many inches every day into streamers two hundred feet long. If it's one of the largest life-forms on the planet, it's one of the simplest too.

Like a plant or a photosynthesizing microbe, a giant kelp makes its own food from sunlight, so in that way at least, it isn't all that simple.

---

[*] See http://www.nationalgeographic.com/animals/mammals/b/bluewhale.

But otherwise it doesn't travel, doesn't change shape or interfere with other life-forms, doesn't perpetuate itself in a series of complicated ways like a malaria parasite, and may not even learn things like a paramecium. Instead, it just makes spores, and beyond that does nothing except wave about as the sea currents move it.

But here I should add a cautionary note—the picture I've just painted of the mindless, do-nothing seaweed is the same picture once painted about most animals and all plants. This was found to be inaccurate and could easily be wrong about seaweeds too.

Obviously, the protists have little in common except in what they're not. Some act like animals, and others act like plants. One of these gave rise to plants, as might have been expected. What's surprising is that one of them gave rise to animals and also to fungi. Who would have dreamed we're related to fungi?

CHAPTER 5

Fungi

To many of us, the word "fungus" implies a mushroom in the woods or an infection such as thrush or itchy vaginal oozing. But this is because most of us don't know the Fungus Kingdom very well. As has been said, the fungus known as yeast is a microscopic organism that consists of a single cell and is to another fungus what a dust mite is to a blue whale, because the size of that other fungus, a famous one in Oregon, is four square miles, or by some accounts, nine square miles, although the latter may be two fungi. Either way, the Oregon fungus must be the biggest life-form on the planet.

Its age is estimated at two thousand to ten thousand years. If it's ten thousand years old, the fungus might have started its growth when a giant sloth was looking at it. We see things like this and we gasp at their longevity. That's because we see ourselves as the norm. Hundreds of life-forms outlive us spectacularly. But even so, it's hard to imagine that an individual who began life with giant sloths is still thriving in Oregon. Yes, the fungus changed by growing bigger, but during all that time, Oregon changed from a prehistoric wilderness to a thriving community with farms, cities, roads, and a human population. Even the kinds of humans—at first Native Americans, then European settlers who at first lived on farms and now live mostly in cities—have changed much more than the fungus.

We're seldom aware of the size of a fungus because most of its body is underground. A fungus starts life as a spore that develops a mesh, a mycelium, composed of filaments known as hyphae,[*] which in function are vaguely like roots. Hyphae are either underground or inside something, and they grow into whatever they're eating to absorb the juices therefrom.

As for mushrooms, we see them as self-sufficient individuals like plants. But a mushroom is to a fungus what an apple is to a tree—a fruiting body. The day-to-day things done by a fungus are done by its hyphae.

We have a divided view of mushrooms. Those we buy in the store are edible delicacies, while those in the woods are feared as poisonous, and many are. I had a close call with such a fungus when a group of us went mushroom hunting in New Hampshire. In the group was an Asian lady who when at home must have eaten silver-colored, dome-shaped mushrooms with little rings of "skin" around the top and bottom of the stem. When we returned from our mushroom hunt, we noticed an *Amanita phalloides*[**] in her basket. These are also called Death Caps, and are so deadly that just a bite of one can kill you, so we shouted that she must get rid of every mushroom in her basket in case they'd brushed against the bad one.

She was shocked. She snatched up the amanita, and since I was the nearest person to her, she shoved it under my nose. "Smell it!" she cried. "Nothing so pure and white can harm you." And she tossed it back in her basket.

She'd pushed the mushroom pretty hard against my nose, and maybe some molecules went in, because later I was in the woods on my hands and knees, vomiting so hard I almost evicted my stomach.

---

[*] *HIGH-fy*—"webs."
[**] *AM-an-EE-ta fa-LOY-deez*. Phalloides means "phallus shaped."

Ghostly pale and shaky, I staggered back to the house where we were staying and, together with some of the others, stole the Asian lady's basket, took it far into the woods, and dumped it. A local animal might recognize an amanita if it was growing, but what if the mushroom was in a pile that looked like food? We crushed the mushrooms and buried them under some stones to prevent a wildlife tragedy.

That lady was the only person whose life I ever helped to save, and it was lucky we spotted the mushroom because we didn't look in everybody's basket, just in hers, and only because she was showing it around. But she was so upset by what we'd done that we felt a little guilty. Even so, we would have felt worse if she'd eaten the amanita. And none of us would be here now if she'd cooked it in the stew we ate that night for dinner.

It seems sad that a fungus would poison us. We share a common ancestor who probably lived in the ocean but maybe in fresh water, which means our common ancestor could swim.

A fungus can swim? That fungus could, and so can some of her descendants. The original swimmers didn't leave much of a fossil record, but one can get a sense of them, because fungi like them still exist. Yes, even today, the spores of certain fungi swim around like fish. The spores of certain chytrid* fungi, for instance, search for frogs and other amphibians to feed on, finding a victim by sensing the proteins and sugars in her skin, then fastening to her as the larva of an animal such as a sponge fastens itself to a rock.

Under her skin, the fungus matures as a cyst, absorbing her juices as a land-based fungus absorbs nutrients from a plant. In time, the mature fungus will produce spores. Some will stay in the victim, and others will swim free to find their own victims. Soon enough, the victim dies of heart failure.

One species of chytrid is believed to be responsible for the decline of amphibians all over the world. How did this happen? There may be

---

* KITE-rid or KAY-trid—"little pot."

several causes, one being that a kind of salamander carries their spores but is immune to their attacks. But our species also plays a role. We humans use frogs in many ways (pregnancy tests being one of them), and the people who collect wild frogs to sell visit all kinds of frog-inhabited ecosystems. An important theory holds that the collectors are spreading the spores of these fungi, which once were not found world-wide. One must hope they don't spread everywhere. To lose the amphibians would be like losing all the birds or all the mammals, including us.

With the exception of microbes and perhaps some protists, fungi may have come on land long before anyone else tried it. How they did it is uncertain—they very seldom fossilize. But photosynthesizing algae were in lakes and ponds, and swimming fungi may have been drawn to them just as they now are drawn to frogs. When a swimming fungus sucks juice from an amphibian, the amphibian eventually dies. But perhaps those algal clumps just made more juice and kept on photosynthesizing, feeding the fungus. If so, this worked out so well for both parties that the arrangement became permanent. Today, this double life-form is called a lichen.

A lichen is two life-forms acting as one. The alga provides food and the fungus provides support and water, which after a rain it can hold for a while. Never think that lichens aren't impressive. I'll have more to say about them later.

Over time, fungi evolved to eat all kinds of foods. Some, like the giant Oregon fungus, are vegetarians and live on the juices of plants. Some others, such as molds, are scavengers who sink their hyphae into dead organisms. And still others are carnivores, who capture live prey. Their victims are usually roundworms, of which there are plenty almost anywhere in the soil, and which the fungus catches with loops of its hyphae. The loops wrap around the worm, capturing it as it tunnels among them, and the hyphae grow into the worm. The fungus can then absorb its juices.

The vegetarian fungi cause damage too, and some may be dangerous, such as the Oregon fungus. Its hyphae penetrate the roots of trees and from there grow up under the bark to absorb nutrients. This weakens young trees and can pose a serious problem for a forest. Or so say some foresters. Some fungi feed minerals to a tree and in return receive some of the sugars which the tree is feeding to itself. Maybe the Oregon fungus does this too. If a fungus sucked juice from young trees for ten thousand years without helping them somehow, would the forest now be the size it is? The trees and the giant fungus know the answer, but do we?

All fungi reproduce with spores, and those who came on land found themselves with serious problems when trying to keep Gaia's reproductive rule. If all their spores fell below their mushrooms, their area would soon be overcrowded. This would be okay for a while and could even have advantages. But it can't go on forever, because sooner or later they'd be interbreeding harmfully, and the food supply around them could fail.

Gaia saw this long ago and designed many ways of non-incestuous reproduction, perhaps most brilliantly with land-based fungi. These became masterful at distributing their spores in remarkable, imaginative ways.

Some use air currents. When the spores are ready, the mushrooms turn their caps upward, almost inside out, like a sweater you grab by the hem to pull off over your head, sending perhaps two billion spores a day into the wind, suggesting that the fungus isn't optimistic about any one spore maturing.

And what if the wind isn't blowing? Some kinds of mushrooms try a little harder by filling their caps with gas and, when it's time for the spores to leave, popping them out and sending them flying. Still other mushrooms give off a fragrance that attracts insects visiting the mushroom, who become dusted with spores and spread them around as they walk or fly away.

Yet perhaps the most remarkable method of spore transportation is found in *Cordyceps*,* the most capable of all the fungi, with multiple species, some of which are medicinal and valued in certain Asian cultures.

Like the plants who attract insects to their pollen, *Cordyceps* fungi attract insects to assist in their reproduction. But unlike bees who willingly visit the flowers that welcome them, the insects helping the *Cordyceps* have no idea what they're doing and wouldn't help them if they did.

Cordyceps have many species and come in different shapes and sizes, but all give out attractive odors to draw their insect of choice. One species uses ants, for instance, another uses caterpillars, and still others use grasshoppers or flies. A *Cordyceps* has little spore-bearing bags that brush against the insect, and when the spores get out, they ooze a biochemical that dissolves a spot on the insect's coating. This makes a hole in the insect where the spores can enter, but also where other things can enter, so the spores produce an antibiotic to prevent infection, an insecticide to prevent attacks from other insects and a fungicide to prevent intrusion by different fungi.

It's hard to know how to view a *Cordyceps* during this part of its journey. Perhaps it's best to view the whole thing in the way we view a malaria parasite—a life-form in a series of shapes and stages. No phase is independent, few if any do more than one thing, and each phase is part of the whole.

When the spores have prepared a victim to their liking, they consume its vital organs while swelling in a yeast-like manner. Then they enter the victim's brain. If the *Cordyceps* is a species that uses ants, the spores cause the ant to climb high up in a tree and bite into the bark or into the large, central vein of a leaf, so her jaws will hold her in place. Imagine being able to manipulate another organism's brain to make that organism do anything, let alone do something as specific as (step one) climb a tree, (step two) find a leaf, and (step three) bite

---

* CORE-diss-eps—"club head."

into its central vein, all of which will lead to its destruction. With all our massive technology and all our studies of brain function, could we manipulate a brain in that manner? Of all the organs in the human body, the brain is the least understood. We're now moving a little in the *Cordyceps* direction—researchers are working on devices that manipulate your brain—but this won't be nearly as complex as what the fungus has been doing for millennia. For all our knowledge and technology, we still have a long way to go.

Not only do the *Cordyceps* manipulate the ant brain, they understand and manipulate the brains of several other kinds of animals, such as caterpillars and flies. If the *Cordyceps* uses caterpillars, it causes the caterpillar not to climb a tree, which might have been her plan, but instead makes her keep walking along the ground, which is not a bad strategy, because without powerful jaws, she would have no way to fasten to a tree (not until it was time to make her cocoon or chrysalis) and would fall off during the next phase of the procedure.

When the victim is secure and in a good place, the *Cordyceps* kills her and sends out a fruiting body to distribute new spores. If the victim is an ant in a tree, the wind carries the new spores away. A caterpillar, in contrast, will have moved a certain distance along the ground, and the spores will disperse where she dies, hopefully in a good place where the young fungi will thrive.

It's unlikely that flies or caterpillars understand the dangers of a *Cordyceps*. But ants do, and if they see a gaunt, unhealthy ant staggering around aimlessly, or up in a tree, hanging by her jaws from a leaf, they carry her far away and dispose of her even if she is still living. The last thing the ants want near their colony is a *Cordyceps*.

CHAPTER 6

Animals

A popular concept holds that plants came before animals. An editor once corrected an article I'd written in which I said that animals came before plants. She thought this was wrong, having learned about such things in college where she'd studied anthropology. So how could she be wrong? If animals came before plants, she said, they would have found nothing to eat.

Ah, but they did. They ate each other. I'm not sure how anthropology addresses the problem, but it stems from the word "animals," which to some means familiar animals such as dogs. It's true that plants came before dogs. But fish are animals. Sponges are animals. Clams are animals. The tiniest insect is an animal. The waters of the world were filled with animals long before there was any such thing as a plant. Therefore, plants will be mentioned in the order of their appearance, which followed the appearance of animals.

After addressing plants in an upcoming chapter, the rest of this book will be devoted to animals, probably because the author is one of them and wrongly and blindly finds the Animal Kingdom to be more important than the rest. However, plants are surely the most important Kingdom, and later I'll discuss them as best I can. But this chapter is about the early types of animals—those that most of us might not recognize as animals—and about some of their abilities, so we can better appreciate what Gaia has done. Our kind of animals, the kind we think of as "higher," are mentioned further on. Even so, we "higher" ones are "higher" only by our own definition, as is only too obvious when we compare ourselves with some of the others whom we see as "lower."

The protist who gave rise to fungi also gave rise to us. We and our fungal cousins then went two ways, although we both were swimmers and both were in the water. Both of us produced a mind-bending array of life-forms, and both of us were busy evolving. But, in certain ways at least, our Kingdom may be more diverse than the others. Some animals are tiny—one soon to be described is less than an eighth of an inch long, and some of the tiniest mites are almost microscopic. Others such as the blue whale are and were gigantic. One dinosaur was a hundred and thirty feet long. Some animals such as *Gastrotrichs**— little aquatic animals that look sort of like worms—live for just a few days, while others, such as deep-water sponges, live for hundreds of years. Some animals breathe air while others breathe water. And some eat plants while others eat other animals or carrion, or eat all three.

Some animals swim, others walk, some walk and swim, others walk and fly. Some animals move throughout their lives, some move only as larvae, some move only as juveniles, and others move only as adults.

Some animals have brains in their heads. Some have brains around their necks. Some don't have brains as we know them but manage without them. Some animals have bone skeletons supporting their bodies, others have stiff chitin enclosing their bodies, and still others have collapsible, unsupported bodies.

Some animals are either male or female. Others are only female. Some animals give birth to live young. Others lay eggs. Some of those

---

* Means "stomach hair."

animals lay eggs on land, others lay them in water, and some keep them in their bodies until the infants hatch. Some animals, after laying eggs, walk off and leave them, while others stay to keep them warm or guard them. And as for the infants, some are larvae that look nothing like the adults, and others are small forms of the adults.

If animals appear in almost every possible format, how are they defined? Back when the life-forms of the earth were said to be either animals or plants, an animal was said to be an organism that moved around on its own, a definition with some merit because no animal is stationary for life. Perceptions have changed, however, and now, as we have seen, many other life-forms such as bacterial "wolf-packs," paramecium protists, and certain larval fungi move quite freely, not because they're washed around in water but because they want to, so to speak. Even plants move because they want to, as I'll mention later.

By now, the Animal Kingdom has no single-cell members, but neither do plants. Animals start life as embryos. But so do plants. Animals have been defined as not having firm cell walls, but neither do amoebas. Animals have been defined by eating other life-forms. But what about the dietary habits of the fungi? With their little hyphae sucking juice from other organisms, don't they do that too? Wouldn't our common ancestor, that swimming protist, have eaten other life-forms just as its descendants, the fungi and the animals, do? And what about the plants called Venus fly-traps? Don't they eat flies?

It seems unclear that the beings in our Kingdom have any unique feature shared with no others, except that some of our members have two matching sides, or in other words are bilateral, meaning that our right and left sides mirror each other except for some internal organs such as the heart. But even this description isn't perfect, because some animals such as starfish, while not bilateral, have radial symmetry in that their bodies are arranged in similar sections like slices in a pie, and others, such as one who will be soon be described, at least have an upper and a lower side.

As for us bilaterals, we're built along the lines of a garden hose. Our bilateral sides, filled with muscles and bones, enclose a tube that's open at both ends and thus is open to the universe—a body-plan that began with a swimming, wormlike ancestor, who took food in at her front end and pushed it through the tube until the waste came out behind her as she swam. It's not a bad idea to organize yourself so that your waste doesn't get in front of you. By no means does every life-form manage this. But it works for us. Our plan was devised by a little ancestor who resembled a hose more closely than most of her descendants do, and whom I'll discuss a little later.

By now, at least two million kinds of animals exist. The first were aquatic and tiny, and some were probably something like the animal called *Trichoplax adhaerens*.* These are perhaps the world's smallest animals, so little-known that they have no common name; so unspecialized that none are males and all are females; and so small that they are barely visible to the naked eye and are best viewed with a microscope. No garden hose here—a *Trichoplax* looks like a tiny, hairy pancake about four one-hundredths of an inch in diameter. She has an upper and lower surface but no front or back end, no left or right side. If she's creeping forward—which she does bacteria-style with her movable hairs—and then wants to go left, she doesn't make a 90° turn as we would do, she just goes sideways, amoeba-style. She has no mouth and no stomach, and is believed to eat like an amoeba, by creeping on top of an alga or a bit of organic debris or even a microbe and squeezing part of her underside around it to form a pouch. Into the pouch she oozes digestive juices, again amoeba-style, and absorbs the nutrients through her outer membrane.

A thing like that is an animal? Like a tiger or an owl? Some scientists don't classify her as such, but others do. Her Phylum, the *Placozoa*, belongs in the Animal Kingdom—thanks to her eukaryote cells, her ability to move herself around, and the fact that her eggs hold embryos. Perhaps she's not as complex as an owl or a tiger, yet she has features that define an animal. Nor is a *Trichoplax* a single

---

* *TRY-ko-plax ad-HEER-ens*—"flat with sticky hair."

cell. Tiny though they are, their bodies have multiple cells. But a *Trichoplax* is so small that with a little imagination she might almost pass for a big microbe.

That every *Trichoplax* is female may not be as strange as it seems. After all, genetic transfer isn't necessary for reproduction. In the procession of life-forms that gave rise to the earliest animals, if any of them transferred genes, they did so only rarely. The confusion here arises because those earlier life-forms didn't have gender. They just split in half microbe-style unless two of them got together and exchanged a chromosome or two. But asexual (non-sexual) reproduction can be found in other kinds of animals such as certain fish and lizards, and is said to have happened in rabbits. Of course, rabbits seldom reproduce like this, but if they do it's because something caused a she-rabbit's reproductive cells to think they had encountered sperm cells. When this happens, the newborn bunnies are females.

After we gave up splitting in half, only one kind of us—the female kind—does the reproducing and keeps the species on the planet. Males were invented later to provide a better way to transfer genes, and sexual pleasure was invented later still, because if an organism requires free will to do something, things go better if it's pleasant.

About seven hundred million years ago, our ancestors were somewhat like the *Trichoplax*. Once there were many species of these little creatures, but today just one remains. Its members occupy most of the world's oceans, so they're common enough. But because they're so small, they were unknown until 1883, when someone noticed a tiny speck creeping up the wall of an aquarium filled with sea water and recognized it as a new, remarkable organism. Controversy was to follow—in 1907 someone decided it might be the larval form of some other animal, a concept that was held until the 1960s, when it was recognized as the adult form of a unique, remarkable, and very primitive species.[*]

---

[*] Personal communication from evolutionary biologist Gary Galbreath.

A *Trichoplax* fulfills Gaia's third requirement—the one for reproduction—in several ways. I know of no other animal who does the same. Like microbes and amoebas, a *Trichoplax* divides herself in two, but she can take this further and divide herself in three. Not only that, but she can grow buds that develop into others like herself. And at the end of her life she makes eggs. She does this if conditions are unfavorable, such as too many *Trichoplaxes* crowded in one place. Her eggs are in the same kind of pouch she uses to digest her food, and when they're ready she swells herself with water and floats upward to expose her underside. There she disintegrates while her offsprings float free. Although all her embryos are females, some of them, once released, may join together and exchange a few genes.

Genetic transfer creates new possibilities for survival, and the life-forms of the earth have been working on it for a long time. By now, genetic transfer has made its way up through the animals who have fancy ways of achieving it, including the pleasure they get. For millions of years, our ancestors trudged forward with this problem until sexual activity gave us the highest kind of pleasure. And coupling for fun placed us humans above those lower critters who couple merely because Gaia said they should.

Or so we believe. Once in Namibia I watched two lions couple, and anyone who thinks that humans get more out of sex than lions do is misinformed. The Namibian lions coupled maybe fifty times without resting, and by "coupling" I mean the whole thing—penetration, humping, climax, and withdrawal.

The male seemed to like it. He'd shut his eyes tight and give a little meow each time he climaxed. But it might not have been so good for the lioness—his penis had barbs to keep him inside her. The barbs collapsed when he penetrated, but when he withdrew they scratched her. Then she'd bellow a deafening roar, swing back her front paw and club him in the face so hard he'd almost fall over.

He seemed to understand. He'd shut his eyes, turn aside, and be polite about it. And a few moments later they'd couple again, so the lioness must have found at least a little pleasure.

At last these lions faced each other, looked deep into each other's eyes, and saw they were sharing a thought. The lioness had rolled on her side when the lion withdrew, so she got up slowly to stand beside him. Then together they walked to a nearby waterhole where they both crouched down and drank. Refreshed, they exchanged another glance. The lion gave a little sigh. Then again quite close together, they walked peacefully back to where they were before and resumed their reproductive duties. How far they had come from the *Trichoplax*.

Perhaps Gaia began to question what she'd done with the *Trichoplax*, giving a perfectly good adult just one chance at having actual infants and losing her life as she does. Was this somehow better than budding or splitting? Did laying eggs and having offspring with little or no genetic transfer get a species anywhere? And why waste a mature, functional animal just for that? Gaia fixed the problem only in some animals, because to this day, a number of animals die soon after reproducing—which says something about Gaia: she has little interest in an individual as long as the species goes on.

Even so, she fiddled around with reproduction for a while, and maybe a million years ago (or maybe longer) she made a jellyfish. These are sometimes called the first animals, but that's debatable, because jellyfish have undergone a considerable amount of evolution since the early days of the *Trichoplax*-types. These jellies reproduce sexually, if without the intense experience of lions—a male jelly coughs sperm cells from his mouth (which also serves as his anus) and the sperm cells drift through the water, perhaps to find a female or perhaps not. And unlike the *Trichoplax* who dies when reproducing, the jellies survive the process. A jelly known as *Turritopsis dohrnii*[*] will literally live forever unless she's eaten by a fish.

T. dohrnii start life as other jellies do, as free-swimming larvae who, in typical manner, attach to rocks on the sea bottom, where each of them becomes a group of polyps that mature to their adult, free-floating medusa forms. As such, they capture and eat other jellyfish. But if

---

[*] *Turry-TOP-sis DOR-nee-eye*—"turret-like being discovered by Dohrn."

one of them grows old, or if she gets hurt, she turns into polyps again. These mature to the adult or medusa state and start the process over. She's still the same jellyfish, just more of her.

Strangely, their remarkable ability has not attracted much attention, as was discussed in an article by Nathaniel Rich in the *New York Times Magazine*[*] about a Japanese scientist, Shin Kubota of Kyoto University, who has studied these jellies for many years, hoping their ability might be applied to humans.

Rich points out that lack of attention led to criticism of the scientific community for not investigating this form of immortality for the benefit of humankind. Many might agree. But the scientific community is smarter than its critics. Aren't there enough of us already? If we all lived forever, would we continue reproducing at the speed we do today? After a few years, we wouldn't find standing room on our planet, let alone food unless we ate one another, so Gaia or the scientists would need to develop an extra-deadly microbe to keep our numbers down.

T. dohrnii is said to be unique in its immortality, though this is a bit of a gray area. The larvae of carrion beetles, for instance, if starved, can revert to an earlier stage of development. Later, when conditions improve, they return to the stage they were in before they reverted, a process they repeat over and over if need be. Isn't that somewhat immortal? What's more, if an organism such as an amoeba divides in half, and the two new amoebas divide in half, going on and on like this perhaps forever, who can say that anyone has died? And how different is an amoeba from an "immortal" jelly who doesn't survive as an individual but as an ongoing group of individuals? And why does it seem bizarre that an animal does this when plants, who send up sprouts from their roots, are doing something like it in our front yards?

In certain ways, and because they seem to keep some bacterial talents, the early animals can seem the most exciting. Most fascinating of all

---

[*] Nathaniel Rich, "Forever and Ever," *New York Times Magazine,* December 2, 2012, 32–39 and 65–70.

are surely the waterbears, also known as tardigrades. I happen to be enthralled by waterbears. I talk about them all the time, and try to mention them in every book I write. I can't always do that, but I certainly can here, and it all began when my dad gave me a binocular microscope with which I looked at swamp water.

I was in college at the time but spent my days in my dormitory room because whatever I saw through the microscope was vastly more interesting than anything said in my classes. One morning I was viewing a bit of swamp water when suddenly a horrible monster loomed up and charged right at me. I was so scared I threw myself backward. My chair tipped over and I fell on the floor. I wanted to get up and run but managed to calm myself—after all, whatever you see through a microscope is sure to be small—so I screwed up my courage and took another look. It was, I later learned, a waterbear.

When I got to know this little creature better—inside as well as outside, because it had transparent skin—I saw round things like eggs in its hindquarters and realized my waterbear might be a girl. And so she was. Most waterbears are female. I would have watched her all day, but I had to attend at least one class because there was a test, so that evening I watched her again. I didn't see her at first, but I moved the slide around until I found her. I got a sewing needle and managed to scoop it under her. I then held the needle up to the light and turned it gently, not seeing her, until at last I barely made out the tiniest speck on one side. The tip of the needle was wider than she was. I put her back on the watery slide. She seemed unfazed by the experience.

She had a nice face. Above her tiny snout, she had two little eye spots. I wondered what she saw with them. If she looked up through the microscope, which she didn't, would she see two monstrous eyes? If I were her, this would disturb me, so I hoped it wasn't possible.

Like me, my waterbear was bilateral. Her body was segmented, vaguely like a caterpillar's, and she had eight legs with little feet, each foot with four claws. She walked as we do, if with more legs, by moving her first and third right legs and her second and fourth left legs simultaneously, then doing the opposite. It seems confusing when described, but it's easy to picture when we step forward on one foot while swinging the opposite arm. Moving alternate limbs keeps you in balance and seems to be a favored way of locomotion among animals. My little captive walked with her head lower than her "shoulders" as bears do, which seems to be why these little creatures are known as waterbears.

Waterbears are in a Genus of their own, with well over a thousand species. They're found in almost every possible damp environment, from wet moss to damp sand dunes, from the Himalayas to deep oceans, and from glaciers to hot springs. But perhaps the most amazing ability of waterbears is that even though they're smaller than the point of a pin, they can withstand five hundred thousand roentgens of radiation, while just one thousand roentgens, or 0.2 percent of that, will sizzle one of us. A dose of more than five roentgens a year is considered too much for an X-ray technician. Even a nuclear blast might not seem like much to a waterbear. They've been on earth for five hundred million years and thus have dodged all the extinctions, including those that cleared away just about everything else.

But perhaps that's not surprising, as waterbears are prepared for any emergency with methods that are somewhat bacterial. If injured, they turn into cysts. Their innards contract, their cuticles thicken, and in this form they repair themselves. If the environment seems too waterbear-unfriendly, a waterbear pulls her head and legs inside her body, which pushes her bodily fluids out. She then looks like a tiny mouse-dropping and is known as a "tun," because whoever called her that thought she looked like a tiny wine barrel. Like a bacterium in endospore mode, a waterbear in the form of a tun can live for many years—some say for a hundred years—during which conditions may improve, or the wind will pick her up and carry her to someplace better.

To think that evolution produced such a marvel! I used to wonder how it happened—how an animal who was advanced enough to be bilateral could have abilities that seemed reserved for microbes. I finally concluded that they might have formed symbiotic relationships with microbes, which they probably ate.

Waterbears don't have intestines, so the food just dissolves inside them, and although their gonads are protected, they're not well protected. It seemed to me that a microbe with some profound ability like the ability to live in acid or a glacier could have found its way into a waterbear's gonads and managed to transfer some genes, after which the hatchlings of that waterbear would have the abilities of both parents, the microbe and the mother. Interestingly, I wasn't the only person to consider this, as a scientific report to that effect was published later. I was ecstatic and filled with pride, but another report then questioned the findings, so they may or may not apply. I hope they do, but what does it matter, except for the value of scientific inquiry? Waterbears are marvelous creatures, whether or not they have microbial genes.

As far as I know, waterbears didn't give rise to anything else. If true, this is tragic, but then, why would they? Change comes about when improvement is needed, and if any life-form needs no improvement, it's the waterbear.

Their different populations now inhabit many kinds of ecosystems. They don't need much food, and the young don't require parental care. My waterbear wasn't going to care for her eggs or even bother to lay them. They'd just stay in her cuticle,* which at some point she'd shed like a snake sheds his skin. As for the need of parental care, even if a predator ate one of those eggs, the hatchling might become a cyst or a tun and wait until it was excreted to resume its more functional form. Or so I imagine. It's true that a waterbear can withstand most hardships, but I didn't want mine to experience a hardship, so I put her back in my jar of water which I emptied into the swamp.

---

* *CUE-tickle*. A cuticle serves the purpose of skin.

CHAPTER 7

Dry Land

Life as we know it began in the ocean, where abundant, very different ecosystems sheltered life-forms of almost every description and still do. "The ocean is the world's largest wilderness," wrote the science writer Sy Montgomery, "covering 70 percent of the surface of the globe. But this vast blue territory is even bigger than it looks from land, or even from space. It's a three-dimensional realm that accounts for more than 95 percent of all livable space on the planet."*

Why would a life-form leave all this to visit a barren, rocky shore? Most of them didn't, so some of those who found themselves on such a shore could have arrived there by accident. For three billion years, we stayed in the water. Each eukaryote Kingdom, whether protist, fungi, animal, or plant, began in the water, to say nothing of microbes. And anyway, where were those shores?

Shores appeared at different times in different places, because the continents were moving and still are. But not quickly. Nature works at both ends of the speed spectrum, at the top of which is the speed of light,

---

* Sy Montgomery, *The Octopus Scientists: Exploring the Mind of a Mollusk* (New York: Houghton Mifflin, 2015), 1.

or 186,000 miles per second. Much slower is the speed of a snail, which a scientist clocked at three inches an hour. And at the bottom of the spectrum are the earth's tectonic plates, which, if they move at all, take about three weeks to creep the width of a human hair. Thus, in an extremely slow process, the early islands—some made from the earth's hot core forcing up lava through the crust at the bottom of the sea—were pushed together into continents that took millions of years to form. One of the earliest, at least in theory, would have been Rodinia ("Motherland") in the southern hemisphere. Estimates vary as to how long it lasted, but that it formed or was forming a billion years ago has been suggested.

If indeed Rodinia existed, and it probably did, it came apart to form other continents—some of them large and most of them scattered. Then about three hundred million years ago, the continents came back together to form the super-continent, Pangaea.* The combining continents didn't know they were meeting one another and kept moving forward with nowhere to go but up, thus forming high mountain ranges around a dry area in Pangaea's middle parts. Pangaea wasn't complete until the mid-Triassic, perhaps seventy-five million years after it started. By then it was the largest landmass in the history of the earth and also the only landmass, covering—and larger than—the part of the globe where the Atlantic Ocean is now, and extending on both sides. The north end touched the North Pole, the south end covered the South Pole, and the thirty million square miles of land in between provided almost every possible climate, with cold areas near the poles, steaming areas in the equatorial coastal regions, moderate areas in between, and a huge, dry area in the middle. This dry area appeared because air masses moving across the vast continent lost their moisture before reaching the inner parts. As such, although the continent went through serious climate changes, Pangaea lasted for a hundred million years.

The tectonic plates are still moving. It's predicted that in the next 250 million years, Antarctica and Australia will join near the South Pole,

---

* *Pan-GEE-ah*—"all lands."

Africa will creep toward the North Pole, and the Americas and Eurasia will creep toward each other until they press Africa between them. The super-continent thus formed will be massive to the north but less massive to the south, where Argentina will join with Thailand, Malaysia, and Indonesia.

The new super-continent already has a name, Pangaea Ultima, and will look like a lopsided donut with a small ocean in the middle. This will not connect with the open ocean, the enormous ocean, the limitless sea that will cover the rest of the world. We don't know who will be there to see this—probably not humans, but waterbears might—and all we can predict is that the surviving species will find huge islands in the ocean made of our discarded plastic, and on land will find fossilized buildings, highways, and cars.

As for the original Pangaea, in time it broke up to form smaller continents. The one called Laurasia eventually became Europe, Asia, and North America. The one called Gondwana became Antarctica, Australia, Africa, and South America. Part of the Gondwana mass came away to form India, so India was off by itself for a while before it crept north to join Asia. Fossils from the original Pangaea are found on these continents, showing that many of us got started there, though not in our present forms. We rode the new continents as they broke away, so that a kind of animal who began in one hemisphere may now be thriving in the other.

The ocean, at least in theory, is accessible everywhere. If you can swim, you can (in theory) go anywhere you like. In reality you can't, because the ocean has ecosystems just like the land, not all of which are friendly to everyone, but even so, the oceans were at least connected, or most of them were for much of the time.

Not so the dry land, with the exception of Pangaea. After life-forms began to live on continents and islands, they couldn't move from one to the other unless they could fly, and for most of the earth's interesting existence, nothing could fly. Such isolation became important. It explains the evolution of many life-forms, and could never have happened without those who make sugar with air and sunlight and let the oxygen float free.

CHAPTER 8

Lichens

So far, we've been considering the various kinds of life-forms from microbes to the earliest kinds of animals, and after this chapter we will move on to plants. But here we must consider the life-form that's none of the above—a small, disregarded life-form that can be found almost everywhere but that few people notice. It's the lichen, and however humble it may be, there's nothing else like it.

Who would believe that something like this is extraordinary? A lichen looks like a little mixture of disorganized moss, quite small, somewhat lumpy, and flat. It might be stuck to a rock or the trunk of a tree, where it just lives peacefully. What makes it extraordinary is that it's composed of two completely different kinds of organisms, a protist and a fungus, bonded together to live as one organism.

As has been said, most of a fungus is composed of fibers that do all the work, and these provide support and also water. When it rains, the fungus collects enough water to last itself and its partner for several months. As for the algae, these are scattered among the fibers. They provide the food by photosynthesis, using sunlight to make sugar from air.

Lichens are so stable that they reproduce as if they were one-occupant organisms. Sometimes they do this by splitting—if a piece breaks off a lichen, it may mature as another lichen. But lichens aren't content with that, so they sometimes reproduce sexually inside themselves. The algal parts and the fungal parts make little bundles of their own genetic material, and when these combine to form a single bundle, sexual reproduction is accomplished, and the bundle pops out to become more lichens.

Today there are many kinds of lichens, found world-wide in moderate or tropical climates, which means that both the occupants of lichens, instead of evolving individually as other life-forms do, evolved while joined together.

They've been around for a very long time. According to a recent study based on DNA evidence, they may have existed for 1.3 billion years.[*] Some evolutionary biologists question this, saying that studies of fossils are more reliable than studies of DNA, and if lichens were present that long ago, they would have survived during a period known as Snowball Earth. It seems unclear when this period started, but seven hundred million years ago it was in progress, waxing and waning, so for long stretches of time the earth would become extremely cold, and most of the water would be frozen.

Snowball Earth isn't said to be one of the great extinction periods, and although dry land was nothing much more than windswept rocks, totally self-sufficient life-forms such as lichens could have been consuming carbon dioxide, which causes warming. If enough of them were present to diminish a warming gas by spoiling carbon dioxide, they could conceivably have contributed to periods of Snowball Earth. Anyway, they were certainly making a little oxygen, which may have been helpful to some of those who followed them on land.

[*] Penn State, "Land Plants and Fungi Changed Earth's Climate, Paving the Way for Explosive Evolution of Land Animals, New Gene Study Suggests," *Penn State Science* and also *Science*, both August 10, 2001.

Today, several animals exist who show how lichens might have started. The most famous of these is the Emerald Elysia, a sea slug. But at least one other slug, some flatworms, and even a salamander put algae to use. These animals live in beds of plankton along with clumps of algae, which they eat. However, although they digest most of an alga, they don't digest its little organelles called plastids, tiny bags full of cells that do the photosynthesizing. Somehow the plastids stay in the animals' bodies and turn them green. Living among green algae as these animals do, the green color serves as camouflage. But there's an extra benefit. The captive plastids, not knowing whose body they inhabit, continue to photosynthesize, feeding the animal who ate them just as they fed the alga who first contained them.

This doesn't seem to be as advanced as the lichens, because an animal with plastids inside it is still an animal, and these animals appeared long after fungi, but their experience could echo the lichen experience. Fungi are famous for attaching themselves to something and consuming its juices, often repaying the donor with much-needed minerals, and primitive, swimming fungi might have penetrated clumps of algae as chytrids penetrate frogs. The chytrid doesn't kill the frog before eating him, she just clings to him while enjoying his juices, and if an early fungus was helping herself to a little juice from some algae, she might have enjoyed it and hoped to stay. To me, this suggests that the partnership could have begun in the water. But perhaps once the partners were formed, they got stranded on a barren expanse of rock where they had little choice except to stay together, needing only sunlight and rain. This is Gaia's Rule Two seen at its most important level, and although it certainly helped the lichens, the oxygen they kept discarding would later help the world.

Modern lichens avoid the exhaust of cars. The trees in my New Hampshire driveway are covered with lichens, as are those on the country road that goes to highway 101. The road doesn't have a lot of traffic but the highway does, and when you're out on 101, you never see a lichen.

CHAPTER 9

Plants

Perhaps the greatest marvel of earth's history, the marvel that changed the dry land from a desolate region of wind, dust, barren rocks, and harmful gasses to a paradise with deep soil, thick vegetation, and fresh air with oxygen, was the evolution of those who could photosynthesize, without whom we couldn't breathe or eat. That we do both is thanks to photosynthesis. As has been said, this was first achieved by a group of bacteria, most especially cyanobacteria, and was later passed on to algae, and later still to plants. Since plants are our most important source of oxygen and also the base of all food chains—without them we couldn't breathe or eat—we should now consider photosynthesis in more detail, starting with sunlight.

To my mind, sunlight is yellow. If you put your hand in a patch of it, you feel nothing except warmth. It has no mass. You can't touch it. I cannot wrap my mind around the so-called color spectrum which ranges from red to ultraviolet, or how, as a plant, you can take it apart, use some to make food, and somehow secure what's left under your surface, where it seems to stay like any other tangible substance such as paint.

A botanist once told me that the green light just reflects off the surface; it doesn't stay there. But if "reflects" meant to him what it means to me—a usage consultant for the *American Heritage Dictionary*—why is a leaf the same color when in sunlight as it is when in the dark?

The answer lies in chlorophyll,* a substance that's able to absorb sunlight and use the other spectrums for energy but not the green. Molecules of chlorophyll are in the little organelles called plastids, mentioned earlier, which plants keep in the cells of their leaves. They find carbon dioxide (CO_2) in the air, take out the carbon (C), mix it with water (H_2O), and discard some of the oxygen (O). Cyanobacteria and photosynthesizing algae do much the same. They float on the surface of a lake or a pond or even the ocean, so they're in sunlight, but they take what they need out of water with a somewhat different result. The sugar made by plants is $C_6H_{12}O_6$, which is dextrose, and judging from the formula, seems to be only half as strong as $C_{12}H_{24}O_{12}$, the sucrose made by the algae.

The early photosynthesizing protists were algae who lived in water, perhaps in shallow pools that sometimes turned to mud or went dry around the edges. Algae like these are still with us, and here confusion arises, because some are now said to be plants. (It's a long story. Some say they're plants, others say they're algae. Lots of species now exist, so some could be plants, and others could be algae, and since it seems uncertain, I'll continue here with algae). Some of their species band together and form the now unwanted mats on the surfaces of lakes. These mats don't seem too different from mosses, the first organisms that could be called plants.

Green algae of various kinds spread around for fifty million years, forming mats that got stuck in wet places such as the banks of ponds and streams. They were sometimes subject to drying, no doubt with natural selection favoring those who could best hold moisture, until about 475 million years ago, by which time some had modified enough to count as moss.

---

* "Chlorophyll" comes from Greek and means "blue-green leaf."

Moss and its immediate relatives are known as bryophytes.* These are liverworts and hornworts. Most still grow in damp places, and instead of roots they have little fibers that enter the soil to keep them in place but don't transport water like roots do. They don't have leaves or seeds like flowering plants who came later—their sunshine receptors are little green twists. As for reproduction, mosses are like lichens and can do this by splitting and also by reproducing sexually. They live in wet places, even in places that are no more than damp, and their tiny sperm cells can swim or perhaps wiggle or drift through dampness, but the result is primitive spores containing what could be called primitive embryos.

We may never know how this came about, but we can get some idea of how early plants could have achieved sexual reproduction from certain kinds of modern liverworts. These liverworts have special, cup-shaped leaves that, when rained upon, make splashes that direct the sperm cells up and out to fertilize the female cells of liverworts nearby, in what may have been an early method.

One achievement of the early plants (and lichens) was to make a little soil, which is composed of sand or clay plus air and some organic matter, mostly from plants who died and fell apart. Today, it takes about a hundred years to make an inch of soil, and this only happens where plants are already growing, shedding their leaves and other body parts. As for lichens, they're mostly on rocks or stuck to trees, but they too could be dropping bits of themselves to contribute.

During the first few million years or so, a few mosses in some damp areas might have made a little soil around where they were growing. Today the dry land is covered heavily with soil. This gives a sense of the time it took to make a world-wide cover. I've seen the process where I live, in a house built by my parents in 1935. Leading to it, they made a walkway of stepping stones with the upper surfaces slightly above ground level. My mind's eye still sees them, and a photo of the

---

* *BRY-oh-fites*—"moss plants."

house still shows them. Today, almost eighty years later, most of the stones are maybe half an inch below ground level. Earthworms helped to push the soil up, but some of its rise was caused by grass making soil in the lawn.

No wonder plants evolved slowly. It took mosses and others 170 million years not only to make some soil, at least where they were growing, but also to evolve the cycads, conifers, and ginkgoes. The flowering plants evolved much later. And again, it took time to solve problems, but those solved by these higher plants allowed them to flourish all over the world in many kinds of climates and without surface water.

All the life-forms of the world came from water, and to this day, none can live without it. But many have found that they needn't be immersed in it, because rain and dew can bring them what they need. Freedom from surface water might have been tempting, because while the plants were diversifying, animals arrived on land. Until that time, the plants had no predators, but the first of the animals lived in pools and streams, from which they would have noticed the nearby plants. While everyone on earth depends on water, not everyone needs to lay their eggs in it or grow to maturity in it, so at some point in their evolutionary history, some animals left their watery homes to nibble the plants that grew near.

These must have been the earliest predators of plants. More were to come, and soon enough, many animals had become vegetarians. Even if the plants were just mosses and ferns, who could resist such easy prey or such bounty?

Anyone who knows where a plant lives can take advantage of it. The plants were forced to reckon with that, so some of them moved away from bodies of water—a step in the direction of developing roots and a kind of tissue known as xylem.[*] The roots were used to find ground water and the xylem was used to transport it through the plant.

---

[*] ZYE-lem—"on wood."

Another step taken by plants was to save what water they had. They covered their leaves and stems with a little wax and covered the rest of their bodies with heavier coatings such as bark. With their bodies protected from water loss, they could live on higher ground.

Even so, the ever-evolving predators had no trouble finding them, and defensive measures were needed. Some plants, for example, modified their sunshine-receptors into needles such as pine needles, a quite brilliant stroke of the conifers, which came to discourage predators. Conifers changed their leaves to needles as an adaption to cold climates, but it can't have escaped them that a predator often prefers a fern or a moss than a bite of something that prickles.

In the manner of poisonous fungi, some plants developed toxins, no doubt for the same reason. An animal browsing on such a plant soon has a bad taste in her mouth and perhaps feels sick. You can see this by watching an antelope, say, browsing on a bush. She nibbles on the east side of the bush for a while, then goes around to the west side, and soon moves on to nibble a more distant bush. That's because the first bush became aware of a predator nibbling and discouraged her with a defensive toxin.

Some trees warn other trees of predators. When such a tree learns that predators are attacking, it not only defends itself with a toxin but also puts out an airborne chemical signal—a pheromone—that communicates with other members of its species. Sensing the pheromone, the other trees prepare themselves for attack by mustering their toxins before the predators find them. Moreover, interestingly enough, it was recently discovered that some trees communicate with other trees by clicking their roots, making sonar vibrations in the soil that communicate with their neighbors, perhaps to discourage them from taking their food or shading others, or perhaps to give warning of a threat.* It's an arboreal form of conversation. Who says plants don't have human characteristics? Anthropodenial is as good as dead.

---

* Monica Gagliano, Stefano Mancuso, and Daniel Robert, "Towards Understanding Plant Bioacoustics," *Trends in Plant Science* 17, no. 6 (2012): 323–25.

For a stationary organism, reproduction is also a problem. Many trees can copy themselves by sending up sprouts from their roots, but is this reproduction or just an interim arrangement? If splitting in half is reproduction for an amoeba, sending up a sprout could be reproduction for a tree. But since the new sprout is still part of the parent and doesn't split off as a new amoeba would, perhaps it isn't. Perhaps the tree that's sending the shoot is just trying to find more sun. Since the tree itself can't move, but its roots are spread widely, it's using them for a secondary purpose.

This certainly happens, and the result can be dramatic. An enormous colony of quaking aspens in Utah was formed by this sprouting process and is said to be eighty thousand years old. Biologically speaking, if not in appearance, the whole thing is the same tree. But sprouting involves no transfer of genes, and this is why trees devised another method. Enter botanical sexual reproduction.

Better to fertilize someone else than to fertilize yourself, which is what happens if your pollen drifts down on your body. The early plants tried to fertilize others, of course—witness the splash-carried liverwort sperm cells—but higher plants took the process further and with more reliable results. Wind, of course, has been pollinating plants for a very long time and still is, as anyone allergic to pollen will testify. But like the liverworts, plants who use wind must make masses of pollen—which is tiny, male sperm cells. Well, sort of—they're not exactly sperm, they're tiny reproductive cells in little containers—all in a cloud that goes where the wind goes, with the goal of reaching another plant's ovaries.

Most pollen doesn't reach its goal—not that the process isn't successful, because it's very successful. But the flowering plants took the process further by enlisting the help of animals. This was brilliant. An animal, perhaps a bee, learns that nectar is deep inside a flower and burrows down to get it. In the process, her fur collects pollen. When she flies away to visit a similar flower, the pollen from the first is scraped off on the next. Many of the little pollen specks will find an ovary, and the resulting offsprings will have two parents.

Millions of years later, our Neolithic ancestors were also to enlist animals for various uses in much the same way as did flowering plants. Wild animals were tempted to visit us for food. We didn't want them to have the food—mostly they stole it—but in time we realized the potential of some of them, and for all we know, plants may have done the same. With a few exceptions, we used animals who walked, while plants used animals who flew. But in both cases the result was similar.

Plants have indentured hundreds of species while we have indentured relatively few, and these include semi-domesticated reindeer and an unlikely kind of cockroach used as pet food. Also, many of us treat animals badly. Plants, in contrast, treat their animal helpers more . . . more what? More humanely? These would include numerous species of bees, wasps, hornets, flies, butterflies, and ants, to name but a few. Each plant attracts its favorites and wants them to be pleased. Some flowering plants use small birds such as hummingbirds, and some even use mammals—mostly little bats, but also other animals such as raccoons and opossums, some of whom are seen with pollen in the fur around their noses and thus are assumed to have been poking into flowers to taste the nectar, just like bees do. Surely no plant depends entirely on possums or raccoons; they just benefit from their occasional visits.

Another important step taken by the higher plants was the development of seeds. Like animals, plants start life as embryos, which are formed in the body of the mother, but after that—again like many animals—the higher plants enclose the embryo in a protective cover. In plants the result is called a seed, and in animals it's called an egg, but except for what's inside these useful coatings, they're the same thing, really—an embryo covered by a protective coat from which, in time, it will either sprout or hatch.

Interestingly, plants and animals developed this feature independently. This seems miraculous. The common ancestor of plants and animals can only have been extremely primitive, probably little more than a collection of eukaryote cells who lived hundreds of millions of years before there was any such thing as an egg, let alone an embryo.

So plants developed embryos in a plant manner and animals did so in an animal manner, a process known as convergent evolution, provided by Gaia when she sees what seems to work.

Such convergence happened elsewhere, if not in the same way. Birds, bats, and insects all fly with wings, for instance, but their wings came independently from totally different ancestors and have very different structures. Whatever far-distant ancestor these animals share lived in the water and certainly didn't have wings.

Even more interesting, however, is that the wings evolved by birds, bats, and insects are all very different in the way they work. Not so the embryos, which in plants and animals are very much the same. Early on, for instance, a vertebrate embryo shows a spine, and a plant embryo shows a stem.

Many plants like to have other plants near them. Trees in a traditional, unmanaged forest profit immensely from the presence of the others with whom they communicate and cooperate. Mother trees feed their nearby children and also other trees of different species, but only the kinds they like, feeding them through their entangled roots. They don't feed the kinds of trees they don't like. Banded together in a forest, trees protect one another from high winds, preserve a moist atmosphere, and communicate through their roots or with airborne pheromones if something bad is happening, such as a parasite attack.

Although plants profit from being close together, little profit comes from constant interbreeding, and like the fungi, plants found ways to avoid this. Some, such as milkweeds, make seeds for wind to carry, but many plants make seeds for animals to carry. A burr, for instance, is a seed with teeth that bite into any passerby who happens to brush against it. A burr is never welcome, so the moment a burr is noticed it gets plucked off and thrown away, hopefully in a place where it will thrive. Several kinds of plants have burrs, some so large that they can puncture a tire. I doubt that any animal eats that kind of burr, but then very few animals eat any kind of burr—an important advantage for the plants who make them.

Some flowering plants took seed transportation to the highest level. Their approach was not aggressive like the burr's, but pleasant and appealing. These plants cover their seeds with edible coatings, often sweet with the sugar they made with photosynthesis. Fruit pulp may have started as a defense against seed-boring insects, in which case the benefit was accidental. But the plants themselves may have realized (in an evolutionary, vegetative manner) that pulp could be more than just an insect deterrent. And not every insect is deterred.

The rest of us cannot resist a sweet-smelling fruit, so whoever finds one eats it, digesting the pulp but not the seeds which not only are small but have resistant casings. The seeds pass undamaged through the gut of their benefactor, who is walking or flying away, and after a day or so will push them out behind him, all packed in a turd which is nourishing fertilizer. There the seed will be ready to grow. Fruit-eaters do this every chance they get and thus could possibly be seen as loosely symbiotic partners with the plants they benefit.

One must wonder about nut trees, however. A nut is an embryo covered with nothing much more than a very hard shell, as is a walnut, or with a not-so-hard shell, as is an acorn. World-wide, the kinds of animals who eat nuts must number in the thousands, and all who do must break open the shell, however difficult. The embryo is chewed up and digested, and no seed planting happens. Not only that, but with such hard shells, how does an embryo manage to get out of the shell to grow? A walnut shell if left on the ground will disintegrate, but this could take a long time, and meanwhile, what if someone sees it? The finder will snatch it up, break it open, and eat the embryo. No doubt the walnut trees chose to take the risk, since most animals who would like to eat a walnut—squirrels, mice, and others—can't break the shell. Animals with powerful molars can, and we humans can. But ancestral walnuts evolved before we appeared to cause them problems, and now we compensate a little by planting walnut trees.

When making shells, the walnut trees and others like them tried harder than the oaks, but oaks are better understood. If acorns are planted by a squirrel who forgets them or is caught by a predator and doesn't live to dig them up, time spent in damp earth rots the somewhat fragile shells quite quickly, and the embryos can start to grow. Squirrels are not the only animals who eat acorns, however. People sometimes eat them but seldom plant them, and dozens of animals eat them with no thought whatsoever that planting them might do some good.

Consider, for instance, the acorn-eating wildlife of New England, ranging from mice and voles to deer and bears, not to mention larval insects. Many of these animals must gain weight in order to survive the winter. Not only do they eat the acorns that lie on the ground, they scrape the ground around the trees in case some acorns are hidden. Under the best of circumstances, many acorns must be planted because only a few will survive, and if too many acorns are eaten, oak reproduction declines. Can the trees do something about this?

They can and they do. Sometimes all the oaks within hundreds of square miles withhold their acorns. A study sponsored by the USDA Forest Service in 2011 shows that all species of oaks do this together. It's a bit unclear just why this happens, although some foresters tell us that it's due to climate conditions. That may very well be so, except that the territories of the oaks who do this—sometimes extending from Canada to the Carolinas, if not farther south—can be subject to different climate conditions, and according to the USDA Forest Service study, the oak trees synchronize their action wherever they are.

Some naturalists have a theory that the Forest Service study might seem to support. Perhaps by communicating with windborne pheromones, the oaks withhold their acorns to let their predators starve. The following year, the oaks will again make acorns, which will then be relatively safe because the few surviving predators can't eat too many, and oak reproduction will revive.

It's interesting to note that some animals use the same strategy as the oaks. (If Gaia has a good idea, she'll use it elsewhere.) To give but one example, large herds of female caribou give birth at the same time. With hundreds of calves born all at once and able to run an hour later, most will survive wolf predation because wolves have territories and keep other wolves from trespassing. As the number of calves increases explosively, the number of wolves stays the same.

In 2008 over much of the eastern United States, the acorn crop failed almost completely. The animals who depended on acorns were desperate. During winter, they starved unless someone fed them, as did this author, who fed and saved twenty-five deer, five or six blue jays, and seventy-five wild turkeys, even though you're not supposed to feed wild animals except for little songbirds. Such is the inconsistency of our species.

I regret my malfeasance, of course, but as an animal myself, I know what hunger feels like, as did those I tried to help. And in a plant/animal conflict I usually side with my fellow animals. I must give warning, though, that feeding deer in winter has serious risks. Just about everyone hates you—if a deer eats your neighbor's ornamental shrub, the neighbor thinks it's your fault for attracting the deer to the neighborhood. But the more important risk is to the deer. Their digestive systems have adjusted to a winter diet, which non-winter foods can disrupt. Deer can be killed by such feeding if it isn't done with information and great care.

I knew the deer as individuals. The following spring, they were alive and well and probably grateful. They came back again when winter started and looked at me through a window to see what I would do about it. And what was that? You'll have to guess. I don't recommend that others do it, but if someone insists, please study the problems carefully first.

As for intelligence, plants don't have brains as we know them. But brains are overrated by our species—witness the brainless "white rat" paramecium mentioned earlier who learned to avoid an electric shock. Like other life-forms without brains, plants know things. They sense sunlight and water and grow parts of themselves to reach these necessities. They sense gravity. That's how they know to grow their roots down and their tops up.

They also sense shadow or relative darkness, as can be seen when a vine climbs a support. The vine knows which side of the support is sunlit and which is in shadow, and grows the cells inside its stem accordingly, not by changing the normal cells that face the support, but by growing extra-long cells behind them. The long cells bend forward over the short cells, thus curving the stem until it leans into the shadow. Meanwhile, the tip of the vine keeps growing normally. Soon enough, the upper leaves sense full sunlight, but the lower leaves still sense the shadow, and the vine bends toward it again. This, when done over and over, wraps the vine around the support.

Brains can cause more trouble than they're worth, leading their owners astray and producing streams of crises, so what's the point of having them if you can solve your problems without them? In fact, growing a line of long cells behind a line of short cells is how plants turn toward whatever they want. But it's dramatic to see a vine do it. And the whole process takes very little time; most plants can turn some of themselves toward the sun in almost no time.

Some plants even walk, as was pointed out in a letter to the editors in *The New Yorker* by Mark W. Moffett from the National Museum of Natural History. Moffett writes, "Certain fig trees can 'walk' on their stilt roots to extract themselves from an overly shady spot, or escape from under a tree that has fallen on them."* Other plants, he tells us, can wander, such as a philodendron that, without changing its length, can move among the branches of a tree until it finds a sunny spot. While it's on the move, its leaves are small and its stem is thin, but when it finds a place that pleases it (my phrase, not Moffett's) its stem thickens, and its leaves grow big. To me, at least, that shows it knows what it's doing.

This and a second letter by biology professor Tobias Baskin of the University of Massachusetts are in response to a fascinating *New Yorker* article, "The Intelligent Plant," by Michael Pollan.** The discoveries reported by Pollan certainly give plants an air of intelligence as with the activities of the philodendron and the other plants mentioned

---

* *The New Yorker*, January 27, 2014, 3.
** *The New Yorker*, December 23 and 30, 2014.

earlier. Baskin feels differently and writes, "Michael Pollan reports on a debate over whether plants have intelligence. They don't."*

Baskin's view is certainly acceptable and has been since the dawn of botany. His opinion is widely held and applies to animals too. But wasn't he practicing anthropodenial? What is intelligence? Something that's only for humans?

According to the *American Heritage Dictionary*, "intelligence" is "the ability to acquire, understand, and use knowledge." No mention here is made of species; no one has said that only people have it. The word "knowledge" might be key here, because when a plant acquires information, this is the basis of knowledge. Then the plant must "understand" that knowledge, plant-style, or it wouldn't be able to use the information it receives. This, then, is intelligence, and I can think of no better word to describe why a vine knows to turn to the shade in order to achieve its goal, or why a fig tree knows where to walk.

And if having intelligence isn't enough, plants also have memory. In a fascinating experiment reported in the fabulous, must-read book, *The Hidden Life of Trees*, by Peter Wohlleben, a mimosa remembered what it had learned about drops of water. Mimosas, says Wohlleben, close their leaves when touched. In this experiment, mimosas closed their leaves when drops of water touched them, but stopped closing their leaves after observing that the drops did no harm. Much later, water was dropped on their leaves again, but the mimosas had learned that the drops wouldn't hurt them and didn't close their leaves, or in other words, they remembered.**

But a mimosa's thought process isn't like ours, and that's what trips us up when we talk about it. We tend to forget that we don't need conscious thoughts for everything we do. No matter what we're thinking, our bodies are working away on their own. Our thoughts can't tell our hearts to beat faster or our noses to stop running. Even if our minds are blank, our hearts will beat, our kidneys will filter, and our lungs will

---

* *The New Yorker*, December 23 and 30, 2014.
** Peter Wohlleben, *The Hidden Life of Trees: What They Feel, How They Communicate* (Vancouver: Greystone Books, 2015), 47–48.

pump. If we're walking on a woodland trail and come face-to-face with a cougar, our eyes fly open, our stomach muscles clench, our knees and elbows bend in a pre-run crouch, and our skin prickles as our body hairs rise, trying to make us look bigger. Our skin doesn't know that it's covered with clothes or that its hairs are tiny. It got a sudden flash that its owner must seem bigger and is doing all it can. Our conscious thoughts play no part here—they kick in later. Meanwhile, the cougar has the same experience and bounds off through the trees.

Plants sense more than light, gravity, and water. They know, for example, about the fungi with whom they have symbiotic relationships, just as their algal ancestors did when they made lichens with fungi. A fungus will line its thread-like hyphae along the root of a plant, perhaps to suck a little juice from the plant but also to provide the plant with minerals. The fungus will receive some juice along with a little payment, because in return for the minerals, the plant provides the fungus with carbohydrate sugars, which a fungus can't make on its own.

And interestingly enough, both the plant and the fungus know if the other is pulling its oar. In 2011, the August 12 edition of *Science* published an account of researchers who found that if a fungus is not providing the right amount of minerals to the plant, or the plant is not providing the right amount of sugars to a fungus, the other will know and may punish its partner by reducing its own contribution. The partner may then reconsider, and perk up its action. *Gosh, what happened to the sugar? Is my partner here upset with me? Was I a little selfish with the minerals?* It's quite something, when you think about it.

Today, almost any kind of wild landscape, from rainforest to tundra, is thick with plants, heavy with plants, because members of their Kingdom are more successful than ours. We celebrate the extraordinary person who lives to be over a hundred, but almost any tree, if in a forest and not damaged, lives much longer. Some black gum trees near my house were growing when Columbus discovered America; many baobab trees in Namibia are at least two thousand years old; a yew tree in Europe is four thousand years old; and an oak in California is five thousand years old. Other kinds of plants have long lives too—the seeds of a flower, *Silene stenophylla*, found in Russia under the Arctic ice—sprouted after thirteen thousand years.

Longevity isn't much use, though, when faced with a species such as ours. Fireballs from outer space, global winters, and volcanic action have been detrimental to plants in the past, but we're doing our part and more to see to their destruction. That a mid-sized species of mammal such as ours could challenge something as widely spread and heavily populated as the Plant Kingdom would seem unlikely. But of the four hundred thousand species of plants now on earth, we humans have managed to threaten almost one third of them with extinction. Even now, many kinds of plants are disappearing. But as for success, we should consider the last group of plants, the grasses. These could be here long after the rest of us are gone, and while we're still here we need them.

After the evolution of flowering plants, about ninety million years would pass before a primitive, rice-type grass appeared, perhaps thanks to animals, probably at first to some of the smaller vegetarian dinosaurs who ate, among other things, little green shoots. Their habits threatened the existing flora, encouraging some to take defensive measures. A grass-like phytolith, which is a bit of silica taken up by a grass as a defense against grazing animals and is deposited post-mortem as a kind of microscopic stone,[*] was found in the fossilized turd of a vegetarian dinosaur who, about seventy million years ago, stopped walking for a moment, stiffened his tail so as not to soil it, bent his knees slightly, and pushed the turd out. Only the turd was fossilized, yet it told us that a kind of grass was present when dinosaurs roamed.

This wasn't the grass that grows on our lawns or fields, although the grass-types were heading in that direction when mammals arose, becoming horse-types, cattle-types, goat-types, and deer- and antelope-types of various sizes. These had wide lower incisors for cutting

---

[*] J. W. Hunt, A. P. Dean, R. E. Webster, G. N. Johnson, and A. R. Ennos, "A Novel Mechanism by Which Silica Defends Grasses against Herbivory," *Annals of Botany* 102, no. 4 (2008): 653–56.

the grass, and large, square molars for grinding it up, and may very well have caused the rice-type plants to evolve as the *Poaceae*,* the modern grasses.

And what a revolution this has been—a plant that grows upward, not from the top as most plants do, but from the bottom and thus can be grazed upon and hardly noticed, a plant that takes root in all kinds of earth as long as there's sun, a plant with tiny seeds that wind can carry, a plant that soon grows back after a fire, a plant that doesn't need much water, and if it doesn't get enough, droops down for a while but stands up again as soon as the next rain comes to refresh it.

Grasses have bypassed the methods used by other plants to reach the sun—the twigs, the trunks, the branches—and left nothing but their leaves and threadlike stems exposed. And the main function of the stems is just to hold the reproductive parts up high where wind can reach them.

All in all, grass has made brilliant arrangements and could be called the most successful of plants. Other plants tried many methods, but grass has outdone them. Grass is the *Homo sapiens* of the Plant Kingdom, but without the same path of destruction.

Grasses became one of the world's most abundant kind of plant, composing one-fifth of the earth's vegetation, covering one-third of the land, and is found almost everywhere, even in Antarctica if not throughout Antarctica. Its multiple species have found ways to occupy a variety of ecosystems, from cities to forests to wetlands to tundra, to say nothing of prairies, savannahs, and plains. Grass has been turned into corn and wheat, into rice, millet, and barley, and is useful to thousands of life-forms—the hundreds of animals who eat it, as well as those who hide in it or make their nests with it, to say nothing of us, who use its species for everything from food and thatches to lawns and ethanol.

After we destroy a forest, grass grows there because grass grows quickly. After the dinosaurs and "dawn horses" and the zebras and real

---

* *Po-AY-see-eye*—"bent ones."

horses, plus the antelopes, deer, buffalo, goats, and many others had grazed upon it, it grew back from the bottom up and is here to this day, triumphant. Certain individual tussocks on the American plains formerly grazed by the buffalo are said to be thousands of years old because, like the mycelium of fungi, the roots, not the grass blades, are the important part of their bodies. The only organisms that really bother grass are bigger plants that shade it. If lightning strikes and starts a fire, the grass and the bigger plants burn together, but rising quickly from the roots, the grass comes back, not always alone, but always first.

Savannah grass has burned since grass evolved, thanks to lightning, but surely was burned more often after our species learned how to set fires. We did this to lure the grazing animals we hunted, after noticing that they liked short grass better than long grass and that after a fire, new grass would come. Antelopes like short grass because it doesn't wave around so it's easy to bite, and also is easy to digest because it hasn't yet grown tough cellulose fibers.

The fires we set would burn for miles and then would go out for some reason—perhaps the wind would blow them back on areas already burned. A few days would pass while our ancestors waited, then little points of green would start to show, and antelopes would be there, munching.

If we could take our planet back in time, that time would be a good choice. If we could keep ourselves there, that would be better. For two hundred thousand years, we were part of an ecosystem as adapted to our species as it was to any other, to the point that many plants adjusted to our grass-fires and even depended on them to germinate.

Today, in several countries with savannahs, setting fires is prohibited by government decree, and the ancient landscapes are pretty much ruined. This is not entirely the fault of the governments, because as our human populations grew, we set too many fires. At any rate, on many an ancient landscape certain bushes—those that fires once kept in check—have taken over and the now-shaded grass is compromised.

Thus, the herbivore populations have changed dramatically, with many of the larger herbivores now gone, especially those that our ancestors hunted. Non-human predators hunted them too, but most of these were shot.

Grass is still here, though. One of the most successful life-forms ever to appear on earth, grass will continue to hold its own even if in small patches. Small patches, that is, while we're around to keep them small. After we're gone, they'll be spreading.

CHAPTER 10

Arthropods on Land

Plants made the dry land habitable. The next to arrive were animals, and these were arthropods. What is an arthropod? "Arthro" means joint, as in "arthritis," and "pod" means foot, as in "podiatrist." The enormous Phylum Arthropoda includes ticks, spiders, scorpions, insects, barnacles, millipedes, centipedes, lobsters, crabs, and others—by far the most plentiful animals on earth today, and also the first to leave the water. They all had legs, although it's not clear why they needed them if they lived in water. Nor is it clear who their ancestor was, although it could have been a sea-going worm.

Considering the benefits of habitable, three-dimensional water when this happened, why an aquatic animal would want to move to land is hard to imagine. But the first who did were not swimming up or down, they were simply walking on the bottom, thus as far as they were concerned, theirs *was* a two-dimensional environment, just one with water instead of air. Maybe that had something to do with their acceptance of dry land. Their jointed legs could move them about, and their chitin skins—the hard, outer coverings that hold arthropods together—kept them from drying.

An early pioneer was a millipede—a heroic individual now known as *Pneumodesmus newmani*.* This millipede perished 428 million years ago to become a fossil in what is now Scotland. But other arthropod pioneers were also trying, if not right with the millipede, then not long after, leaving some of their fossilized bodies as well as their little fossilized tracks on fossilized shores. These pioneers surely came from the sea floor where others like them live today, walking around to scavenge food that drifts down from above. Lobster-types were also among the first to try land. One of these was discovered as a fossil in China, in a place that was then a shallow sea.

Because the sea was shallow, an outgoing tide could travel a long way—too far and too fast for a sea-going arthropod to keep up. If she was anywhere near the beach when the tide went out, she would be stranded. But a chitin-covered animal with legs and feet could manage, and as for breathing, anyone who plans to cook a lobster will see that she can live a long time on whatever air she finds in the refrigerator. Natural selection would have favored those who could wait for the tide to come back, giving them a chance to develop the capacity for long-term air-breathing. These creatures, who during that time were designed for eating motionless aquatic morsels, could have found the abundance of motionless shoreline plants inviting. After all, the early plants needed direct contact with surface moisture, and many of them surely lived right next to water or even partly in the water—maybe in swamps or on the banks of ponds and streams. Such conditions must have been encouraging to the early land arthropods.

Once on land, the millipedes are thought to have eaten mosses, not only because mosses grow in wet places where there wasn't much else to eat, but also because some modern millipedes still eat mosses. The pioneer millipede probably did, as there was little else to nourish him. But to think that his preference lasted 428 million years! Gaia made a winner when she made that millipede. He and his descendants

---

* *NEW-mo-DESS-muss new-MAN-ee*—"bound lung discovered by (or named for) Newman."

weathered the climate changes, the drifting continents, the volcanic eruptions, the comet strikes, the asteroid strikes, and the terrible mass extinctions—and as long as they could find a little moss, they managed.

The same cannot be said for every arthropod who left the water. Given a new environment with few if any predators and unlimited possibilities, other arthropod pioneers evolved in many different directions and occupied many ecological niches as evolution directed them. In time, groups of them became important features of these ecosystems. They interacted profoundly with the ever-evolving plants, not only by eating them (thus stimulating them to protect themselves), but also by forming symbiotic partnerships with them, carrying their pollen in exchange for a little nectar, until the plants and the arthropods were deeply entwined with each other.

Today, more animals are arthropods than anything else. Not only did they branch out widely, but some did what members of dominant taxonomic families tend to do and what vertebrates did later—grow to enormous sizes. An insect like a dragonfly appeared who weighed about a pound and had a wingspan of over two feet—perhaps the largest insect ever to appear and surely the largest dragonfly-type known from any time. Even today, modern dragonflies are among the largest insects. But for truly large size, we return to the millipedes, as a millipede-type appeared who was eight feet long, probably the biggest land-based arthropod ever. We can't say the millipedes didn't evolve at all, because they did, but only to be different sizes and different types, like centipedes. Most giant arthropods have gone extinct, which may be just as well. Even so, a millipede in East Africa today is fifteen inches long, and if he walks across your foot he looks even bigger.

Over the millennia, the land-based arthropods kept their exoskeletons and flourished. Those who stayed in the ocean did too, and some, such as certain crabs, joined their land-based relatives by finding homes in fresh water. Today, fresh water crabs have branched into more than a thousand species, but their achievement pales beside the land-based arthropods in general—the crab-types (4,000 species), the millipedes and centipedes (20,000 species), the wasps, bees, and hornets (22,000 species), the ants and termites (26,000 species), the spiders and scorpions (44,000 species), the butterflies and moths (175,000 species), the flies and mosquitoes (240,000 species), and the beetles (400,000 species), which account for 30 percent of all the animal species on our planet. Compare these with us, the Hominids (5 species—orangutans, gorillas, chimpanzees, humans, and bonobos), and consider that all this came from a handful of pioneers who may not have chosen to leave the water but found that the water left them. Of course, the arthropods had more time to produce so many species, and today are the most abundant animals ever. They're found all over the world.

CHAPTER 11

Vertebrates

Our Phylum, the vertebrates, has not been as prolific as the arthropods. But who can blame us? We haven't been around as long. The significant part of our development took place in the water, which we didn't leave until our important physical qualities were established. Our ancestor is believed to have been a sea-squirt, either a tunicate or closely related to a tunicate, an aquatic life-form that, like a sponge or coral polyp, stays fastened to an undersea support for life. Today, many such species exist, and a good example might be the little bluebell tunicate that looks like a flower growing from a rock. Many tunicates look something like that, if not as pretty, down in the ocean, sessile* for life, filtering the water for food drifting by. That's fine if you've found a good place to stay, but if your rock gets crowded or your species is solitary, your infants can't fasten beside you. So how do they find a new place? They do it in the way of sponges and coral polyps, starting life as larvae who drift around or swim around until they find somewhere

---

* This wonderful word, "sessile," is science-speak for "not free-moving" or "permanently attached" and is mostly used for sea creatures such as tunicates.

to fasten, there to mature to adult form. It seems that such a larva was our ancestor.

But here's where some scientists diverge. Did we really descend from the larva of a sessile filter-feeder—a little larva who found a rock and started to fasten, then imagined that the sessile life was not for her and swam away to change the world? Or was she a little worm-like swimmer to start with and later evolved in two directions, one to become a sessile filter-feeder, and the other to become a vertebrate? It's hard to know which version is more likely. Adult free-swimming tunicates exist today and are found in all oceans. But in either case, a little swimmer can be victimized by ocean currents, and to help with that problem, our ancestor evolved a notochord, a flexible rod running from one end of an animal to the other, providing support where muscles can attach.

Often a nerve cord lies above it, as it does with the tunicates. The device helps with swimming and orienting, because with extra strength and control, a swimmer can readily move in any direction, perhaps finding the right rock in less time, perhaps exposed to less risk than it would be without a notochord. Once fastened to a rock, a tunicate no longer needs the notochord. She's still a larva, if already maturing, so most of her notochord is still in her tail, and when she feels secure on her rock, she brings the tail around to the front of her body and absorbs it, notochord and all.

But our ancestor matured as a swimmer, not a sessile tunicate, and thus kept the notochord for life. Unlikely as it seems, these little swimmers gave rise to fish, amphibians, reptiles, birds, and mammals. Like larval tunicates, we all start life with a notochord, and interestingly enough, like adult tunicates, we lose it. But we still need its service, so we revise it along with its nerve to become our spinal cord and column.

Named for our spinal cords, we are Phylum Chordata. Although there's not much resemblance today between an astronaut walking on the moon and an extinct filter-feeding sea-squirt stuck to a rock on the bottom of the sea, to honor our ancestors for what they've done, we include the tunicates in our Phylum. So here we are with our spinal cords, reading books, driving cars, waging wars, and looking through microscopes at tunicate fossils.

As far as we humans are concerned, we got our start when vertebrates left the water. Enter the amphibians, which today are the frog- and salamander-types, as well as a few amphibians known as caecilians,* who look like large worms. These are found in tropical areas, where they live underground and are seldom seen. And they lack a common name—at least in English. It may be hard to imagine that we belong to a clade that includes not only Darwin, Beethoven, and Einstein, but also obscure worms in burrows, even though, like all amphibians, those worms can think, even if not as well as Einstein. We all descended from a larval sea-squirt who kept swimming.

But we didn't diversify quickly. We were fish before we were amphibians, and for our evolution to get started, we had to leave the water. This happened as fish were diversifying. Some had a bony joint, which in us might resemble a shoulder, but in fish it served as an anchor for their side fins, giving stiffness to the fins and strengthening the fish in water.

Not just any fish could have entered the land masses. Some of them managed by moving into streams and swamps. Fish who hung around the mouths of rivers—perhaps hoping for plant matter or dead arthropods to wash downstream—could have grown tolerant of partly fresh water and moved upstream, especially into the swamps—first, perhaps to swamps beside the sea, and then to swamps farther up the rivers. A swamp with lots of plants would have attracted the early arthropods, and these would be food for venturesome fish.

But swamps are subject to drying. Just as the sea can move back from a shore, a swamp can shrink so that only mud remains. Not only that, but swamp plants grow in thick clumps that slow down the flow of water, meanwhile dropping their leaves and other parts that, in time, decay. Slow-moving water plus organic decay causes stagnation,

---

* *Sess-SILL-yans*—"blind creatures."

leaving whatever water is left with little if any oxygen, and making gill-breathing difficult for fish.

Certain fish would do better than others in coping with the problem. Although fish get their oxygen from water, which they gulp and force through their gills, some fish, such as lungfish, must gulp some air from time to time or they'll suffocate. A lungfish-type would have had more luck surviving in a stagnating swamp.

Some fish have six fins. But the lungfish-types have four fins, and all descendants of that fish have four legs or adaptions thereof, such as two legs and two arms or two wings. It gives us the name "tetrapods," which is Latin for "four feet" and is here an unavoidable instance of science-speak because there seems to be no synonym. Even so, it's important, because when you think about it, those who descended from that fish are the only animals with so few legs. Most animals, if they have legs at all, have more than four, from millipede-types who have hundreds of legs to spiders who have eight legs and insects who have six legs or six legs plus wings. It would seem that in all creation, our kind has the fewest number of legs, leading one to conclude that if a lungfish-type wasn't our ancestor, it's hard to imagine who was.

However, the fins of most lungfish may not have been as strong as those that appeared on the fossil of a lungfish called *Tiktaalic roseae*.[*] His fossil was 375 million years old and was found by paleontologists Neil Shubin and Ted Daeschler on Ellesmere Island. Shubin described the influence of *T. roseae* on our evolution in his popular book, *Your Inner Fish*.[**]

T. roseae was long and rather thin. Unlike some other fish, he had a neck, shoulders, and a pair of primitive lungs. And his fins were stronger than the fins of other fish. If water was too shallow for *T. roseae* to swim, he could have waded. Perhaps this fish, who could get along in the air, at least for a while, began going on land to catch the early arthropods he kept seeing around the plants. The timing coincides

---

[*] "Large rosy fish," from Inuit.
[**] Neil Shubin, *Your Inner Fish: A Journey into the 3.5-Billion-Year History of the Human Body* (New York: Pantheon Books, 2008).

quite well—plenty of arthropods were present on land. Maybe they were too tempting. Maybe *T. roseae* waded out of the water on his stiff fins to hunt them.

It's true that other fish exist today who can walk on land and even climb a tree. They're still fish, though, not amphibians. Again, the problem is their habitat—most of them live in mangrove swamps where water stagnates, so they often do better when out in the air. If their skins are wet, oxygen comes through them. In science-speak, the word for this is "cutaneous respiration," and in regular English it's "skin-breathing." If these fish are on land, their front fins are strong enough to keep them upright if they want to look around. When doing this, they look like seals propped up on their fin-like arms. These fish are known as "amphibious fish," and the mudskipper is a good example. I saw a video of a mudskipper out of water, braced up high on his front fins with his mouth open wide. He is shouting[*] at someone, perhaps another mudskipper. He looks so brave and determined, so sure that his shouts are daunting, that he almost seems like a bold little mammal. And yet, for all their ability, mudskippers are not thought to be our ancestors. Nor could they have been, as our line was already on its way. I say this with some regret, as I would have liked them to be my ancestors.

---

[*] Believe it or not, many kinds of fish make vocal sounds.

CHAPTER 12

Amphibians

Just because a fish could breathe a little air, why would he want to spend time on land when the water was filled with more nutritional opportunities, kept its temperature relatively stable, and was a three-dimensional ecosystem where he could move in any direction including way up or way down? No one can say for sure, of course, but predators may have played a role. While you were swimming around in search of a smaller victim, someone bigger was swimming around in search of you. It's interesting to note that the likely vertebrate pioneers were small compared to other aquatic residents, and thus would have been likely victims.

During the millions of years of aquatic evolution, predators of all kinds and all sizes were appearing in the water. But the only predators on land were arthropods, and few were big enough to cause a problem for anything the size of a fish. Maybe that inspired the fish-types to try life on dry land. If, as an ever-evolving fish, you could search for food without predators trying to catch you, you could expect to live long enough to produce abundant offsprings and fill the earth with others like yourself, which the new land-based arrivals then did. This was an unexpected benefit and fortunate too, because life on land had other kinds of problems that took thousands of years for this new kind of animal to solve.

The new land animals who once were fish were something like the mosses. "Amphibian" means "double life." They found a way to live on land but didn't free themselves from water. They lay their eggs in water, and during their larval stage are water-breathing swimmers with gills and fins—even with dorsal fins that keep them from rolling sideways, and lateral lines to tell them how deep they are in water. The above describes a fish and also a larval amphibian. Only as adults do amphibians change their fish-forms to adapt for life on land, or at least for visits to the land, by breathing air in various ways—some not only through their noses but also through their skins—and changing their fins into legs.

Like the fish who gave rise to them, early amphibians had scales which would have hindered skin-breathing and contributed to amphibians evolving in two directions. One direction led to the modern amphibians who needed more air and learned to skin-breathe, and the other led to reptiles, a substantially different kind of animal, reasonably independent from water and well equipped for life on land.

As for amphibians, they arrived after the arthropods had been spreading over the land for sixty-five million years, adapting to occupy the habitable ecosystems created by the plants. The arthropods supplied the oncoming amphibians with interesting food, to the point that even today, no adult amphibian eats anything other than arthropods— the one and only source of their diets.

After becoming part-time land animals, the amphibians did what the arthropods did, which was to fit into different kinds of ecosystems, if always wet or moist ones, and over time they developed a range of different shapes and sizes. Some were enormous, such as a giant known as *Prionosuchus*,* who lived 270 million years ago and

---

* *Pry-ON-oh-SOO-koos*—"saw crocodile."

looked like an alligator. He was thirty feet long and weighed about two thousand pounds.

An amphibian at two thousand pounds? That's big for an alligator and outlandish for an amphibian, certainly by modern standards. *Prionosuchus* was the largest amphibian now known. His fossil tells us that he was an adult, or certainly not a larva, and his size tells us that he took the evolutionary path taken by many another groups of animals, which was to become enormous. Fossils of sharks were found with his fossil, suggesting that his evolutionary path had taken him back to sea. But he was so big that maybe the sharks didn't bother him.

The ancestor of the non-reptile types—the frogs, salamanders, and caecilians—took a different turn. Most of the descendants remained fairly small, the exceptions being two kinds of giant salamanders, one in China who is almost six feet long, and the other in Japan, about five feet long. Both salamanders are facing extinction because people eat them or sell them as curiosities.

Modern amphibians have lungs that a mammal would find inadequate. But all of them breathe through their skins as well as their noses, and all but a few depend heavily on moisture. The exceptions include a desert frog who found a way to rid his body of urea with almost no water. But most amphibians live in damp areas or next to ponds where they can get in the water, and if they must travel, perhaps to find mates or else a better pond, they travel at night when the air has more moisture.

Then too, they have problems with temperature. The temperature of water changes slowly, and when they were fish they could rise up or sink down to find a climate they liked. Out on two-dimensional dry land, they had no such benefit and were forced to cope, often by moving only after the sun had warmed their bodies to a functional state, and by burying themselves in something while they waited. Yet for all their disadvantages, they managed very well, and with all those arthropods to feed them, they ruled the earth for fifty million years. It's hard to imagine amphibians dominating anything for so much time. Fifty million years is five hundred thousand centuries.

Today three kinds of amphibians are left: the frogs and toads, the salamanders and newts, and the caecilians—those reclusive, wormlike creatures who live in burrows and may never be known as one of Gaia's triumphs. The three groups have evolved about 7,500 species (the estimates vary somewhat), with the caecilians providing about 200 surviving species, the newts and salamanders about 680, and the frogs and toads about 6,500, which today amounts to 85 percent of all amphibian species—and thus is huge.

But compare this to the numbers of surviving arthropod species—the flies and mosquitos with 240,000 species and the beetles with 400,000 species or vastly more than all modern amphibians put together and perhaps more than all amphibians ever, so one might imagine that the arrival of amphibians did little to impact the arthropods no matter how many they ate.

Then too, a mass extinction occurred. It began perhaps 372 million years ago* during the Devonian, during which many kinds of amphibians perished, giving them only eighteen million years of evolution as opposed to eighty-five million years for the arthropods. So this too may have played a role. Interestingly, the Devonian extinction had less of an effect on arthropods in general and almost none on insects.

As for frogs, it's no wonder they're so successful, at least for their taxonomic class. Their survival skills are wide-ranging: to name but a few, some are poisonous and thus don't appeal to predators; some are tiny and therefore inconspicuous; and some are enormous, or at least very big for a frog. The biggest frog today—the Goliath frog of equatorial Africa—is a foot long and weighs eight pounds.

The frogs we call toads have hardened their skins, which modifies their need of water. And at least one kind of frog, the Darwin's frog of South America (so called because Darwin discovered the species), has departed from the age-old custom of laying eggs in the water and instead lays them on damp ground. The father frog stays near the eggs until the tadpoles start moving inside them. He then takes the eggs

---

* Several dates are offered, but all are in that era.

in his mouth and swallows them into his vocal sack—the big balloon that puffs out under a frog's chin when he's singing. He keeps them there from the time they hatch as tadpoles until they morph into their adult frog form. Surely he feeds them while they're in his vocal sack. (That's quite an ability to swallow food not only into your stomach but also into your vocal sack.) But how does he know when to let the little frogs out? What must it be like to be a father frog with a collection of infants under your chin for whom you are responsible? As tadpoles, the infants would be squeezing around together. Would their father then begin to feel their new little feet poking and scratching, thus signaling that his charges are in adult form? Or would he know to let them out after a certain amount of time? Or would the little ones come out on their own?

Interestingly, all that's left of the amphibians today are those with permeable skins and the need to be constantly moist. They certainly make the best of it, but the form with scales didn't vanish. An early form, known vaguely as "reptilomorphs" of the clade Reptilomorpha, kept their scales, which helped them modify in different directions. Most had the original body design (a small head, a neck, a longish body with four legs and a tail—nothing like a frog with a big head and no tail), and also waterproof skins. I wasn't familiar with a name applied to them—"reptilian"—which according to the dictionary means "of or related to reptiles."

My research seemed inadequate. I turned to the internet and learned from various postings that the first reptilian came down from the sky and wrote the Bible. I hesitate to question the research of others and was impressed by one researcher's confidence. "Open your eyes to the truth," he says. Even so, I became suspicious. I'd thought that writing was invented later, and I'd also assumed that the reptile-types had toes rather than thumbs and fingers and probably couldn't hold a pencil long enough to write more than a word or two, much less something as long as the Bible. So I gave up on everything "reptilian" and focused on "tetrapods that evolved from amphibians." And what a revelation it was! They invented one of the most important evolutionary steps ever taken!

Enter the amniotic egg. Now we can answer the question of which came first, the chicken or the egg. The amniotic egg came millions of years before the chicken and didn't need to be in water. Its leathery covering was thicker and stronger than the outer membrane of an amphibian's egg, with tiny pores to let in oxygenated air for the fetus inside, wrapped in membranes for its feeding and hydration. The leather-like cover became the shells of bird and dinosaur eggs, and the membranes, forms of which were kept by all future egg-layers of this lineage, became placentas.

Various life-forms, including reptiles and mammals, evolved their own kinds of placentas, but the idea of a placenta in its basic form—the membrane around a fetus to keep it fed and wet—was a huge step forward, at least for land-based egg-layers. It gave reptiles and mammals the name of "amniote," which is also the name of the abovementioned membrane, and it comes from Ancient Greek. It was the name of the bowl that caught the blood of a sacrificial lamb.

The first of the amniotes were just one step beyond the amphibians. To think they could have invented something so important! And because the amniotic egg is used by all kinds of animals, not just people, it's been even more useful than the Bible!

Amphibian eggs were mercilessly eaten by fish—and still are, with the result that many amphibians such as toads and salamanders have learned that their eggs are safer in vernal pools, if they can find them. These pools are in the woods, filled with spring rains and melted snow, without access to brooks and streams and the fish that come with them. But vernal pools form only in certain kinds of hardened soils and thus are relatively uncommon.

One imagines a slowly emptying vernal pool with amphibian eggs inside it. By the time it's almost dry, most of the eggs are shriveled. But some of them with slightly harder skins are still okay, and from these hatch little survivors. Gaia sees this, and knows that soon enough, or in thousands of years by our standards, those who can lay such eggs will replace their relatives who can't.

But even these land-based eggs weren't safe. What could be more tempting than a clutch of helpless eggs, out in plain sight where anyone can see them? They don't even know you're looking at them. That's why most animals who lay eggs on land either protect them or hide them, usually by burying them, often with leaf-litter, a custom that must have been invented early on.

The amniotes skipped the larval stage entirely and hatched in the form of little grownups, thus leaving amphibians as the only vertebrates with two life stages. Eggs that didn't need to be in water, and then the protected skins and half-decent lungs of infants who could soon care for themselves—developments along the lines of higher plants—fortified their owners for life away from open water.

While these achievements were unfolding, the amniotes were taking two directions: one toward us, the mammals, and the other toward dinosaurs, pterosaurs, modern reptiles, and birds. As for the early reptile-types, some things didn't change. As arthropods keep an all-insect diet, so do reptiles keep an all-animal diet. Of the ten thousand kinds of reptiles today, only a few, such as iguanas and some tortoises, eat plants.

CHAPTER 13

Proto-Mammals

Much of this next stage took place between 310 and 320 million years ago on Pangaea. We know this because fossils affirming a residence on Pangaea have since appeared world-wide. For instance, fossils of a marvelous mid-sized animal in the mammal-line called *Lystrosaurus*[*] were found in South America, South Africa, China, Mongolia, India, and Antarctica. Fossils of certain reptile-types were found in Africa, India, and Antarctica. And fossils of a certain fern were found in all of these regions. Those fossils were on Pangaea when it came apart.

By then the amniotes had divided into two important groups, both of which were critically important. The earliest group gave rise to the mammals and from there to us, and the second group, which arose much later, gave rise to everything else—dinosaurs, pterosaurs, crocodiles, modern reptiles, and birds.

We all know about dinosaurs and find it hard to believe that our lineage appeared long before theirs. This seems especially hard to believe because many early members of our lineage could easily be mistaken

---

[*] *LISS-tro-SORE-us*—"shovel lizard."

for scary-looking dinosaurs, at least in appearance if not always in size. Some were about ten feet long with large heads and mouthfuls of teeth that were different shapes—somewhat along the lines of our teeth but not like a lizard's teeth, which are all pretty much the same. Other characteristics also point in our direction, showing previews of our lineage, but few of us have heard of the animals who evolved these characteristics and fewer still can name them. Unlike the name "dinosaur," which everyone knows, our ancestors' name is reserved for the PhD programs.

The name is "synapsid." It means "fused arch" and refers to the bone bridge that begins on the side of your face by your temple and is made by your cheek bones and your upper jaw. If you slide your tongue behind your upper teeth, you will feel your temporal arch, and you can look back 315 million years to an animal who looked vaguely like a little lizard (but wasn't a lizard) who had the first version. Millions of years would pass before that arch was perfected, and I'm not sure what was so wonderful about it that kept it going for 315 million years, because otherwise its owners evolved in all directions. But it certainly seems to work, therefore it began as an improvement. After all, it produced a shelf with holes on the sides for big bite-muscles to fasten. Thus it helped us bite hard—a feature quite lacking in amphibians—and then kept going. Gaia herself may have such an arch. But then, maybe she doesn't. Non-synapsids such as turtles and crocodiles don't have such an arch and some bite very hard without it.

Synapsids began on Pangaea, adjusting to a variety of climates and spreading to dry places where amphibians couldn't go. An ice age was fading, so although the southernmost part of Pangaea might have been frozen, the interior areas might have been something like the modern African savannahs—dry, but not true deserts—getting drier toward the massive continent's center, while tropical forests with tree ferns and cycads grew along the coasts.

Thus a variety of ecosystems were in place, and in many, the synapsids did well. They evolved in different directions—some as herbivores, others as carnivores, and some no doubt as scavengers. Some had huge sails on their backs, possibly for keeping cool. A sail draws your body heat up, and its wide, thin surface gets rid of it quickly. A sail could also help warm you if you stood in the sun, so its purpose may have been an early stab at endothermy, which is science-speak for self-warming.

Then too, a sail makes you look bigger when seen from the side, so perhaps sails were meant to make their owners seem scary. Or perhaps they attracted the opposite sex and were used for reproductive purposes. Later, some dinosaurs had sails. In fact, quite a few animals had sails during that period, but whether for regulating temperature, or for making you appear large to your predators, or for courtship purposes, these sails may not have worked as well as Gaia had hoped. After all, when seen from the front, a sail just looks like a thin piece of plastic and isn't very scary. Today only a few small lizards have them, so sails almost seem to have been a prehistoric fashion, or perhaps were an interim device, to be replaced when better things evolved.

Some synapsids remained small, and some became enormous. Some normally walked on all fours but could also walk on their hind legs. Most who walked on all fours did so crocodile-style, with their knees bent and their elbows out.

As time passed, however, some species left fossilized tracks that were revealing. These tracks show five toes, the first and fifth toes short, the second and fourth toes longer, and the third or middle toe longest of all. If you look at your hand you'll see the pattern. This means that pressure on the foot came in the middle, which means that the foot was under the body rather than beside it. The feet that left the tracks were on straight legs and thus were like those of many modern mammals, including us. It also means that our hands have kept the synapsid pattern, and that synapsids invented a modern way of walking.

They may also have invented urine. Our tetrapod ancestors who had moved away from water still needed water in their bodies, and had to conserve it as best they could. So they rid themselves of urea (which, if dry, can look like tiny little pale-colored crumbles) by adding small amounts of water and mixing the paste with fecal matter. This took place in the cloaca. "Cloaca" is Latin for "sewer," and is the last part of its owner's intestine. Everything that leaves the owner's body leaves from there, from bodily wastes and informative odors (if the owner makes scent-marks), to sperm from males and fertilized eggs from females.

Waste from a cloaca is soft but semi-solid, something like yogurt. A bird's dropping is an example—a totally different phenomenon from the watery stream of urine and the solid scat that we mammals produce from two different openings. This was an advantage, because our guts are loaded with microbes, and having a separate exit for the turd protects the urinary tract from infection. Did early synapsids arrange this for us? It might seem that they did. But urine doesn't fossilize, nor do bladders or urethras, and although we can guess, we may never know exactly how we came to have two different openings, or how synapsids peed.

The synapsids probably didn't have scales, but because skins also don't fossilize, little is known about their skins. However, it's been suggested that our ancestral synapsids may have had glands under their skins that produced liquids. No one knows how many glands, or if the liquid they produced became the ancestor of sweat. But because it was a bodily fluid, it could have had nutrients in it, which means it could also have been the ancestor of milk.

Our kind was on its way. But when we see skeletons of early synapsids in museums, or photos of their skeletons on the internet or in books, we look at them in awe, never thinking to look at our hands or push our tongues against our upper teeth and feel the fused arch they gave us, or taste our cuspids, our canine teeth, and wonder how we got them. Instead we see scary dinosaur-types with huge jaws that make us uneasy, and we're thankful that these critters aren't here now.

But they are here now. They're us, and we're not that different from them. We should think of Richard Dawkins's famous image: you are standing by your mother, holding her hand. She's holding her mother's hand, who in turn holds her mother's hand, on and on, back through time, until the hand being held is a chimpanzee's. Dawkins stops here, but we can go much farther, on and on with the hand-holding mothers until we come to an early synapsid who has a fused arch and looks something like a reptile but with different kinds of teeth. She didn't have hands. We'd hold one of her front feet.

Synapsids ruled the world for eighty million years. That's eight hundred thousand centuries and they're ruling again, since we mammals belong to their clade. But damage to their nearly endless reign took place during the great Permian-Triassic or P-T extinction, which began about 250 million years ago and was by far the worst the world has yet experienced. Its progress was complicated and still isn't fully understood, but a fairly basic theory holds that it went on for over a million years as lava kept pouring from the Siberian Traps, which are fissures in the earth's crust under what is now Siberia. The oceans heated to about 100°F, and enough greenhouse gasses appeared to turn the earth into an oven.

It's said that 86 percent of all Genera were lost. This included 96 percent of marine life, 70 percent of land vertebrates and also many insects—the only massive extinction of insects—because plants were hit hard along with everything else. There wasn't much food of any kind and not much air worth breathing.

The other name for this frightful event, the "Great Dying," may not seem harsh enough. Biodiversity was so seriously compromised that maybe fifteen million years went by before conditions stabilized and life-forms returned in numbers sufficient enough to be seen as normal.

As for the synapsids, after ruling the earth for eighty million years, their numbers sank from multiple species to relatively few. Among those who lived was a platypus-type who became our ancestor. "Platypus" is Latin for "flat foot," and there's more to say about her later.

CHAPTER 14

Dinosaurs

During the long reign of the synapsids, another reptile-type was evolving. These were known as archosaurs.* They were distinguished, among other features, by their teeth, which were set in deep sockets, and they started the clade which today includes crocodiles, pterosaurs, and birds. But these are just the tag ends of a huge, extremely successful group that includes the thousands of different kinds of dinosaurs who appeared after the Great Dying. At the time of the extinction, many were small and needed less of everything, including water. Small size has helped many species during the various extinctions.

What the early dinosaurs were like may be seen in a little one called *Nyasasaurus* (lizard from Nyasa), who lived about forty-five million years ago, was three feet tall and ten feet long from his nose to the tip of his tail, and weighed about 130 pounds. Smaller dinosaurs also appeared later, such as a little carnivore called *Microraptor* ("little one who seizes"). From his nose to the end of his tail he was sixteen inches long. He had feathers and four wings like an insect and could glide

---

* *ARR-ko-sores*—"ruler lizards."

but probably couldn't fly. He walked on his hind legs, leaning forward turkey-style, and he probably caught and ate small lizards.

Another small dinosaur was *Xixianykus*,* who was twenty inches long and walked on all fours. Fossils of other small dinosaurs were uncovered, but if only one kind of each is found, it's hard to tell if the owner was a grownup. By the time they appeared, the giant dinosaurs were emerging, given millions of years and many kinds of ecosystems such as those on Pangaea. Dinosaurs replaced synapsids as rulers of dry land and went on as such for 145 million years. Some dinosaurs, those now known as avian dinosaurs, continue to this day as birds.

It helps to look at the past with this in mind—the highly successful, widely spread synapsids remaining on earth as mammals, and the magnificent array of dinosaur species—estimated to be in the multiple thousands although many more will probably be discovered—remaining on earth as birds.

Ever since someone invented the shovel, people have been digging up dinosaur fossils. The small fossils, if found, got little attention, and the big ones were thought to be the remains of giants or dragons. Scientific studies of dinosaur fossils began in the 1800s. What was said to be the biggest dinosaur was found in 1877—an herbivore called *Amphicoelias fragillimus*.** He was estimated to be almost 190 feet long and weighed about 135 tons. This was based on a hip bone, a leg bone, and two vertebrae, but if I have it right, the leg bone was lost or crumbled, and the written description of the fossil had some problems, so it's now said there was no *A. fragillimus*.

Now sometimes billed as the biggest dinosaur is another plant-eating quadruped named *Dreadnaughtus* ("fear nothing"). His fossil was carefully unearthed between 2005 and 2009 and showed him to be about fifty feet long and almost thirty feet tall. He is thought to have weighed perhaps fifteen tons, and although there are other contenders, he may prove to be the biggest of all dinosaurs so far. We sometimes

---

* ZIX-ee-EN-ay-kus—"claw from Xixia."
** Am-fiss-EE-lee-us fra-JILL-im-us—"cavity on both sides, very fragile."

think of the carnivorous dinosaurs as big, and so they were by our standards. But the biggest dinosaurs were herbivores.

As for the carnivores, fossils of *Tyrannosaurus rex* ("tyrant lizard king") were found in 1902. He was judged to be forty feet long and weigh seven tons, and soon became the most famous dinosaur. His image has since been challenged by another monster called *Spinosaurus* ("spine lizard"), who walked on his hind legs like *T. rex*. He may have been sixty feet long, thus bigger than *Tyrannosaurus*, but he seriously lacks the fame.

Sharing the fame was *Brontosaurus*, the "thunder lizard." Most of us know of her too. Brontosaurs were said to be so heavy they had to stand in ponds where the water could help support their weight—a theory based on whales, equally enormous but too heavy to support their weight on land.

I was a child when this theory was offered and, like most children of the time, I was fascinated by dinosaurs but didn't think a brontosaur amounted to much except in size, because an animal too fat to stand was not inspiring. We children used to wonder how the brontosaurs ate.

We were taught (correctly) that they were herbivores, but if they had to stand in a pond to keep upright, it wouldn't be long before they ate all the leaves they could reach, and then what? How did they find more food without leaving their ponds and walking to others? Did they crawl? Wouldn't the tyrannosaurs kill them if they found them struggling on the ground? Most of our dinosaur questions were never answered, including this one, so we just grew up in ignorance.

Fame and confusion have haunted the brontosaurs. Eventually their name became invalid. Doubts arose about the brontosaur and a similar but smaller dinosaur named *Apatosaurus* ("deceptive lizard"), who had been discovered earlier. It began to seem that the apatosaur was a young brontosaur, and if a life-form gets named twice, the scientists use the first name given, so the name "*Brontosaurus*" was changed to "*Apatosaurus.*"

Then the fossil specimens were re-examined, only to show, in 2015, that these in fact were two different but related dinosaurs, as was at first supposed. Meanwhile, studies had continued, showing that neither one of them needed to stand in water or even live near water. Both had legs as big and strong as trees and were themselves so big and strong that few other dinosaurs would have bothered them. If a predator, however large, had been brave enough to tackle such a monster, the huge intended victim could easily have defended herself by the method of a modern iguana who, if confronted with a menace, lashes the offender with her tail.

Once I was struck on the leg by a lashing iguana—a little female who misunderstood my actions and made an understandable mistake. She didn't just wave her tail, she twisted her hips and arched her whole body. The blow was incredibly strong, and my leg hurt for half an hour, during which I thought of a brontosaur lashing. Thousands of times bigger than the little iguana, she would have shattered her predator. It seemed to me that being hit side-on by an eighteen-wheeler at highway speed might do the same kind of damage, so I'm guessing that most predators left the brontosaurs alone. Time and again, large size and great strength have proved their value, as these qualities seem to have done for the brontosaur, a life-form that survived four hundred and twenty times longer than we humans have been on earth, and surely we will never do as well.

Yet fantasies about the dinosaurs go on. Since any large carnivore is usually portrayed in some pre-attack position, it's common to view the world during those 145 million years as an ongoing dinosaur war zone with the quadruped herbivores as clumsy, mindless victims.

Did this start with the name of *Tyrannosaurus*, the Tyrant Lizard? It hasn't stopped. Two recently discovered tyrannosaur-types were named *Teratophoneus* ("monstrous murderer"), and *Lythronax* ("king of gore").

That's how we see dinosaurs. We never imagine a *T. rex* as a newly hatched infant, fluffy and unsure as she views the world of the Jurassic for the first time. Her view might include her siblings struggling out of their eggs, and possibly her giant mother who is looking on with some anxiety because she has so many little infants and some are starting to wander off. Can she protect them all? Just because tyrannosaurs were giant apex predators doesn't mean they didn't have feelings, so to assume that their feelings were rage and greed seems a pity. But it's probably more exciting than assuming they felt hope of better weather or relief that their infants were accounted for.

When my brother and I were children, the rage/greed theory was widely accepted, if not by most scientists, then certainly by us. Fascinated by dinosaurs, we would imagine ourselves in the Jurassic, encountering what we thought was a typical dinosaur. To this day, I see a *T. rex* striding turkey-style toward us, and we're not much taller than his ankles. Then he'd be over us. We'd see the dark cave of his huge open mouth coming down at us. At this point, we'd think of a way to escape, but this was unrealistic.

If our fantasy continued in a more likely manner, his enormous, dagger-sharp teeth would be on both sides of us. For a horrible moment he would chomp us, but then he might see something else of interest—his rival in the distance, maybe—so he would quickly gulp us down to deal with whatever he was looking at, and gosh, we'd be in his stomach. We'd be badly wounded with our knees squeezed up under our chins, and digestive juices would be dribbling on us.

However, our fantasy never went that far. If it had, we would have faced the obvious, with ourselves in his cloaca, which we would exit in a dropping that looked something like a turkey's but much larger, with little chunks of fabric from our clothes sticking out of it. We preferred to see ourselves outrunning him, dodging to confuse him, and because he was stupid, or so we believed, he wouldn't know where we'd gone. Then we'd gloat because of our triumph. We were smaller than him, we'd tell ourselves. But being humans, we were smarter.

Many theories have appeared about dinosaurs. I learned of one, not widely supported, that suggests they didn't vocalize, based on the fact that their descendants, the birds, vocalize with a syrinx and their cousins, the crocodiles, vocalize with a larynx. The syrinx and larynx are very different organs formed on different evolutionary paths, and because such organs don't fossilize, no evidence exists that dinosaurs had either one. Nor does evidence exist that they didn't. For all we know, they could talk.

Most scientists reject the no-voice theory. One of these is Mark Witton, the author of *Pterosaurs*, who points out that dinosaurs "almost certainly chirped, hissed, and bellowed because living reptiles do." He says "The real question is, when did dinosaurs develop the capacity to make more bird-like noises—when did they start to honk, squawk or sing?"* In my mind, the question then would be with what organ, a syrinx or a larynx? I'd bet a dollar to a dime it was a syrinx, which they then passed on to birds, or some of them did. Given time, maybe some kinds of dinosaurs had a syrinx and other kinds had a larynx.

When one considers what vocal sounds are for, the assumption that they vocalized makes far more sense. Consider the scream, for example. Most land vertebrates scream. We do it for several reasons, the most important being that we're scared. Perhaps we're surprised by some danger. We don't plan to scream—a noise just flies from our mouth.

Let's say a predator springs on us from behind. We scream. Perhaps a friend or relative will hear us, understand that we're in trouble, and rush to our aid. More likely, another predator hears us, realizes that someone is becoming a victim and hurries to the scene, hoping to obtain the victim for himself. Our predator will be distracted by the newcomer, and in the scuffle that follows, we may escape. It doesn't always work that way, of course, but because screaming is all but universal, it must be worth trying.

Then too, a scream is made for every kind of listener, so the screams of different species sound more or less alike. If you knew for sure what kind of animal was screaming, your response might be selective, and that defeats the purpose. Perhaps that's why some animals adjust their voices to scream within the common sound range. For instance, a bat,

---

* Mark Witton, personal communication.

who uses ultrasound, lowers her voice to scream, and an elephant, who uses infrasound, raises her voice. Thus their screams are heard by the widest possible audiences. Because a scream gets everyone's attention, the more attention you get, the better your chances of help.

I'd say the scream was one of Gaia's best inventions and that during the 145 million years the dinosaurs ruled the earth, they must have screamed too.

They also may have made infrasonic calls.* These travel far without fading, and many kinds of big animals, including the dinosaurs, are and were social and needed to keep in touch with their groups, however far apart.

The puzzling theories present an unrealistic picture of dinosaur science. They abound on the internet, though. Another one, inspired, it seems, by a computer, is that the big quadruped brontosaur-types with long necks couldn't lift their heads but could only move them sideways. Perhaps to challenge this theory, another interested person with a computer "proved" that giraffes couldn't lift their heads either.

But not all theories seem peculiar, and most address fascinating issues. For instance, how big were infant dinosaurs when they hatched, and how big were the eggs, especially the eggs of big dinosaurs?

Or could dinosaurs manage their body temperatures? Could they raise their temperatures from the inside (endothermy) or were they warmed by the sun (ectothermy) like reptiles and amphibians?

A third question, if with little input from scientists, would be were the dinosaurs smart? The question is hard to answer because thoughts don't fossilize, but one can look around at other life-forms and make an educated guess.

A fourth question, and in some ways the most important, would be what did dinosaurs eat? A dinosaur's teeth show whether her diet

---

* These are calls below the range of human hearing. Elephants make calls that people can't hear at all. Hippos and lions and others, probably as yet to be discovered, make calls with intrasonic components.

was vegetables or meat. But could the giant herbivorous dinosaurs maintain so much weight by eating leaves?

As for the first question, the one about dinosaur eggs and infants, an internet site with a flair for the obvious tells us that "dinosaur eggs are eggs laid by dinosaurs." That's a start, but it wasn't enough for the scientists Louis Chiappe, Lowell Dingus, and Rodolfo Coria, who found a nesting area made by *Saltasaurus* ("horned lizards from Salta")—quadruped herbivores of the brontosaur/apatosaur-type with small heads, long necks, long tails, and heavy bodies. The scientists found that several hundred saltasaur females, perhaps all together, perhaps a few at a time, visited a nesting site in what is now Patagonia, where each of them scraped out a large, round hole with her hind feet. She then laid about twenty-five eggs in the hole and covered them with leaf-litter and dirt.

The eggs had hard shells, like those of all dinosaurs and birds, and were round, about seven inches in diameter, with the capacity of about one gallon. The infant, when hatched, would have weighed about ten pounds. Thus he wasn't much heavier than a newborn human infant who, let's say, could weigh about eight pounds like mine did. But my infants were less than eighteen inches long and grew to be five to six feet long, weighing on average about a hundred and forty pounds, while a saltasaur infant grew to be forty feet long and weighed eight tons. This is because my infants stopped growing when they reached a certain age, while it's been suggested that dinosaurs grew throughout their lives. This is hard to determine, however, because like all who live in the natural world, most dinosaurs didn't live to full maturity. This is determined by their fossils. The skeleton of a fully mature dinosaur, like the skeletons of birds and mammals, has a final, polished layer of bone which says to the world, "I've stopped growing." Most dinosaur fossils don't show this. On the other hand, many modern animals, including turtles and crocodiles, grow throughout their lives, or grow until they get too big for their internal organs to support them.

Where did the idea to keep growing come from? From an evolutionary point of view, was it new? Or was the new idea to stop growing?

Both seem to have found evolutionary success, as both are still practiced, if by different life-forms, but is one form better than the other? A search of reliable sources suggests that nobody knows.

The second question, the one about dinosaurs and thermal regulation, contains a question of its own, because in some ways warm-blooded animals seem to gain advantage over cold-blooded ones. Of course, in other ways they don't—warm-blooded animals such as birds and mammals need to eat more often than the cold-blooded reptiles do. Even so, it's been suggested that some synapsids who survived the great P-T extinction had become or were becoming warm-blooded. So if this is true, how did dinosaurs overwhelm our ancestral synapsids if the dinosaurs weren't self-warming too?

Of course, the dinosaurs didn't really overwhelm the synapsids—the extinction did that—but as to the possible advantage, assuming there is an advantage, one theory points to the sails of dinosaurs who had them. A sail might warm its owner if he stood in the sun, but most dinosaurs didn't have sails, and besides, according to a favored theory, sails were for cooling. However, dinosaurs had air sacs in their bodies and hollow bones like birds do, and this also concerns self-warming. When a bird inhales, some oxygen goes into her bones and the air sacs, then is carried by the bloodstream to her lungs, so she gets more oxygen than we do when we take a breath. Bird-breathing is the most efficient form of animal respiration, and perhaps was true of some dinosaurs too. This could be a sign of self-warming, which requires more oxygen than sun-warming.

Many dinosaurs had feathers or fuzz, perhaps including *Tyrannosaurus rex*, who may have had fuzz only while young, which again is a sign of self-warming. Fuzz, feathers, and fur keep heat stored in the body and block unwanted heat from coming in. Thus a body covered with fibers stays at a more even temperature than does a body with naked skin. However, the evolution of dinosaur skin has proved to be complex, and a recent theory suggests that the tyrannosaurs may have lost their fuzz or feathers and replaced them with scales. The skins of some dinosaurs fossilized—these were *hadrosaurs* (something like tyrannosaurs) and *ceratopsids* (something like *Triceratops*)—and these dinosaurs had scales.

As for fuzz or feathers, the smaller you are the more helpful these are, because, as Bergman's law points out, the smaller you are the more surface you have in proportion to your mass, and since you lose heat through your surface, a bulky body is less vulnerable to the cold. So it could be that dinosaurs when small had fuzz or feathers and lost these coatings when they grew big. Probably the enormous herbivore dinosaurs had no fuzz or feathers—as adults, these monsters would have held heat for a very long time and would have had difficulty cooling even with bare skins. Their modern counterparts are elephants. Adult elephants have bare skins and huge, thin ears—a mammal's version of sails—all for heat reduction.

Although the question may never be answered, and plenty of arguments exist either way, it's now thought that many dinosaurs were on the way to self-warming and some may have achieved it, at least to some degree. They share a common ancestor with crocodiles, who started to be self-warming and later returned to sun-warming, which shows that self-warming can come and go, and whatever works best for a life-form at the time will be favored by natural selection.

And then there were the pterosaurs—another issue of the archosaurs. Although totally unknown to many of us, the pterosaurs were fabulous animals who coexisted with dinosaurs and were almost certainly self-warming. These marvelous creatures require a chapter of their own, following this one.

Were dinosaurs intelligent? For a while they were thought not to be because, unlike us with large brains in proportion to our bodies, dinosaur brains were small. The concept doesn't mean much now, though, because if a paramecium can learn, brain size is no longer diagnostic. Even so, the general assumption was that dinosaurs ran pretty much on automatic pilot. But this was believed to be true of all animals. Many pet owners would disagree, but if scientists learn something in graduate school that's later rejected, some of them retain it anyway.

"At times it seems as though some scientists are the ones with conditioned reflexes," writes Nicholas Dodman, a professor of veterinary science who studies animal minds. "[These scientists] hold stubbornly to their mistaken beliefs, all evidence to the contrary."*

We're not good at measuring intelligence. For instance, we measure other life-forms against our children. I heard a scientist say on television that a Giant Pacific octopus "has the intelligence of a four-year-old child." Many people make such comparisons, believing that they're easy for non-scientists to understand. But we non-scientists are smarter than is sometimes supposed, and even we can see that the measurement seems invalid. Certain mollusks (the clam, oyster, snail, squid, cuttlefish, and octopus group) can solve difficult, man-made puzzles that would stump members of the Mensa Society. For instance, from high on a shelf, an octopus in a tank watched other octopuses trying to find their way through a maze. When his turn came, he'd seen enough to whisk right through the maze, although by then he was seeing the maze from within while before he was seeing it from above. I doubt that adult humans could do that. Four-year-old children certainly couldn't.

We never think to measure a human by a young animal's standards, and wisely so. The man who wins the twenty-meter sprint might be equal to a kitten. If we measured a rabbit's intelligence using a horse as a standard, would it prove something?

As for intelligent dinosaurs, we should look at what's around us. From fish to mammals, the list of intellectual abilities is mind-bending. Even insects have consciousness. They feel fear, for instance, and they remember things. It's hard to imagine that a hundred-foot-long, thirty-ton dinosaur (*Argentinosaurus*—"Argentine lizard") wouldn't do at least as well as a tomato hornworm, one of whom was given an electric shock when she smelled a certain odor and continued to avoid that odor even after she became a moth with a totally different body and an utterly different survival strategy.

---

* Nicholas Dodman, BVMS, DACVB, *Pets on the Couch* (New York: Atria Books, 2016), 11.

It's been shown that some fish learn more quickly than people. Some can recognize themselves in a mirror.* A Border Collie memorized the names of a thousand objects.** I'd challenge any human to name a thousand objects. And Irene Pepperberg, a scientist who studies animal minds, tells of her famous parrot, Alex, who grasped the concept of "nothing."***

This was impressive, if you think about it, because in the natural world there's no such thing as "nothing." When you visit a bush to pick the berries only to find that birds have eaten them, you might say you found nothing, but that just means you didn't find what you wanted. The bush is still there, and trees and grass are all around it, not to mention the sun, the clouds, and the sky. Even so, a parrot envisioned the entirely human, unnatural concept of "nothing."

Alex did this by recognizing that there was no difference between two objects. When asked what was the difference, after a moment's thought he said "None." That's very far from saying "the same" or even "not different"—phrases which Alex could have spoken—and may sound easy to us, but we're used to such things and we grew up with such ideas. For someone who hasn't, it's an enormous concept.

Were dinosaurs intelligent? Parrots are dinosaurs, and some of the smarter mollusk-types, such as the octopus who analyzed the maze, have been around for five hundred million years and coexisted with them.

No instincts exist to address every problem, especially the multitudes of problems that surely appeared during those 145 million years of dinosaur times. Thought and memory work much better. Many dinosaurs were social, living in large, multi-generational groups with the need to interpret one another. Some hunted cooperatively and would need quick, coordinated reactions. All of them knew what kinds of foods to eat and where to find them, and some went on

---

* "Manta Rays Recognize Themselves in a Mirror," *New Scientist*, March 22, 2016.
** John W. Pilley Jr., *Chaser: Unlocking the Genius of the Dog Who Knows a Thousand Words* (Boston: Mariner Books, 2014).
*** Irene Pepperberg, *Alex and Me* (New York: Harper Collins, 2009); John Pickrell, "Parrot Prodigy May Grasp the Concept of Zero," *National Geographic*, July 15, 2005.

long migrations and would need to know the way—it's one thing to migrate as a bird and follow the stars, and another thing to migrate on foot through miles of uncertain, irregular terrain that's subject to annual changes. In other animals, ourselves included, such things require a certain amount of consciousness, including memory and planning. Non-avian dinosaurs were on earth longer than almost any other group of vertebrates, and avian dinosaurs are still here as birds. It would come as no surprise to learn that dinosaurs had consciousness and plenty of it. The only surprise, and a big one too, would be to learn that they didn't.

As for the fourth and possibly the most significant question—what dinosaurs ate—alas, it's been said that *T. rex* was a scavenger. This theory—widely publicized but attributed to just one scientist—came as a shock to *T. Rex* fans because we see him as a hunter. Why did he have huge jaws and a powerful body if he didn't need to kill his victims? And if someone else took the trouble to kill a victim, why wouldn't that someone eat the best parts? Yes, I know that lots of animals scavenge—they're gourmands, not gourmets—and even the most committed carnivores don't always turn away a little carrion if they come upon it, but as for me, I'll never view the theory that my hero *T. rex* was a scavenger as anything but a bad guess.

The real mystery, then, is that the biggest dinosaurs ate plants. Could animals weighing fifty tons get enough energy from leaves? Eating would have taken all their time, and it would seem miraculous if a plant were left alive during the Jurassic.

A fascinating paper, "*In Vitro* Digestibility of Fern and Gymnosperm Foliage: Implications for Sauropod Feeding Ecology and Diet Selection,"* addresses the question. The authors, Hummel et al., were able to find the right kind of nutrients in a number of plants, such as gingkoes, that not only grew in dinosaur times but are still here today and available for study.

---

* Jörgen Hummel, Carole T. Gee, Karl-Heinz Södekum, Martin P. Sander, Gunther Nogge, and Marcus Clauss, "*In Vitro* Digestibility of Fern and Gymnosperm Foliage: Implications for Sauropod Feeding Ecology and Diet Selection," *Proceedings of the Royal Society B* 275 (2008): 1015–21.

The research yielded answers that make one think that the huge, herbivorous dinosaurs were selective and didn't waste their time eating foods with low nutrition. They seem to have favored trees such as umbrella pines with large cones that hold plenty of nourishment. Some of these trees grow very tall with branches only at the top, forming an umbrella-like dome with cones on its branches and bunches of nutritious needles. Not just any dinosaur but only a very big dinosaur with a long neck could reach them.

This also suggests that herbivorous dinosaurs could probably recognize plants like a botanist and remember where they grew, so if these herbivores denuded a grove, they'd know how long it would be before they could go back there. Such dinosaurs might live for many years and thus could amass a large amount of such knowledge. Their tracks were found near the trees that fed them, and these grew in groves along rivers.

When my brother and I were children, we imagined ourselves attacked by a tyrannosaur. Now I'm older with a different mind-set, and I imagine a grove of trees beside a river. The image is entirely my invention—Hummel and his colleagues must not be held responsible because their paper reports the nutritional qualities of different kinds of plants, not imaginary pictures of those who ate them.

That was left for someone such as me, so I'm sitting on a rock at the edge of a river with tangles of giant ferns and groves of cycad trees along the banks. Around me are mosses and ferns that blend into bushes, with a tangle of tall trees behind them, some about sixty feet tall, some without branches for the first forty feet. The drooping branches bristle with clusters of large cones and thick needles and a breeze is moving gently.

The river is narrow but with a smooth current. It's mid-afternoon, and I'm on the east side facing south, listening to insects humming in the bushes and the soft sound of the water, where I see a little creature swimming. It's dark colored and swims quite fast, trailing a V of ripples. Soon enough it hurries up the bank and disappears. It went into a burrow.

Then I hear a voice from far away, a deep, throbbing call, almost buzzing, coming from somewhere in the rolling land beyond the river. The call goes on and on, as if whoever is calling has enormous lungs that hold a huge amount of breath. I listen carefully because the call is faint. At last it stops. I wait.

Then I hear something moving slowly through the woods across the river. The throbbing call came from the south, but the rustling sound comes from the west. Something huge is making it, coming slowly but purposefully toward the river, never pausing, as if whatever makes the sound has a goal in mind. At last, through a narrow space high among the trees I get a glimpse of something gray and curved, like the top of a thick gray pipe bent downward. It's moving.

Moments later, an enormous oval head pushes up through the trees and turns in my direction. I'd been looking at the neck, which was long. The head is also long and smooth with no nose, just nostrils, and no visible ears, just dents on both sides. The skin is gray with black bands from the nostrils up around the eyes and over the top of the head. The eyes seem flat, and they don't seem to move in their sockets. They are pointed at me but they belong to a dinosaur, I realize, who surely sees me sitting here but doesn't seem to care.

I'm surprised. But then, why would I interest someone like that? An elephant might not pay attention to a rat, however unfamiliar, nor might a blue whale care about a mackerel. I was nothing.

At the river's edge, the enormous creature exhales loudly and lowers its head to drink. Then it wades ponderously into the river. The creature is so big that its front feet are stepping through the water while its hind feet are still walking through the woods. But soon its rear comes out of the woods with its long tail lifted. You'd think a tail as heavy as that that would drag, but no. Its owner keeps it stiff so it curves up slightly.

Three other dinosaurs follow, one almost as big as the first, the other two smaller. I remember that among many kinds of herbivorous animals, groups of females are common and groups of males are uncommon, so I assume these dinosaurs are females with a reason to be together. They seem to move slowly, but they cover ground quickly. I can hear their breath, like soft wind gusting, and the suck of their footsteps as they walk up the bank in my direction. All four of them look straight ahead, not at me or at each other.

We hear the throbbing call again, faint and far, a voice from the edge of the world. The dinosaurs don't seem to notice, but could it be one of their kind? At last the smaller dinosaurs stop walking and turn their heads as if to listen, although the two others move on. They seem to have no interest in whoever is calling.

I am, though. Is it a predator? If so, I want to go home. But the dinosaurs here seem unconcerned. They're too big to worry, and whoever is calling is far away. Then too, although I know nothing about dinosaurs, I do know that predators are quiet while they're hunting. Whoever is calling wants to be known.

The dinosaurs who stopped must have seen me, but if they did, I didn't interest them either. They walk up the bank at normal speed, their huge legs wet and shining. They're leaving tracks. In the wet earth, the tracks are deep, but higher up where the earth is damp, the tracks are shallow. The smaller dinosaurs mixed their tracks together so I can't make them out, but the big dinosaur made a separate set of tracks. I see the huge print of her hind foot quite far behind the print of her front foot. Her hind foot is flat and has three huge toes, but the front foot, shaped like the slice of an enormous pear, stem forward, has just one toe. I'd never seen a track like that. I hadn't thought it possible to have just one toe. But is it? I'd like to see her feet, but now I can't. She's in thick vegetation.

Now the group is eating. The two biggest ones with the longest necks can eat from the tops of the umbrella pines, where they bite off the cones and chew them. The smallest dinosaur rises up on her hind legs, bracing with her tail as a squirrel might, to reach a cone, then drops back down to chew it. I see her throat move as she swallows. She rises up again and bites the branch that held the cone, dragging her teeth down the length of it and stripping off the needles. The branch flips back up, almost bare of needles.

Maybe she's satisfied with just the needles. Even so, I wonder why she doesn't eat another cone. But the next time she rises up to reach one, the biggest dinosaur moves up to her and eases her aside. The smaller dinosaur goes to another pine, bites a branch, and again strips off the needles. Perhaps the biggest dinosaur wants to be the one who eats the cones.

As they move around the grove, the big dinosaur comes near me. I'm somewhat worried. If I seem like nothing, she might step on me. But I'm afraid to stand up and catch her attention. Suddenly she lowers her head. She seemed far away before she did this, but when her monstrous head swings toward me, her nose is only ten feet away. She's so big she blocks the view. I can't see beyond her. A storm-like wind, warm and sweet-scented, blows at me from her nostrils.

I panic and start to scuttle sideways, but the huge head swings back up. I see the underside of her chin. She lowers her head again and swings it sideways. The underside of her chin looks like a low-flying plane. She must be looking at something beyond me. I still don't interest her. But I keep moving sideways as carefully as possible, my heart beating like a jack-hammer as I go because I imagine her picking me up like an insect just to see if I'm edible, and I don't stop until she turns away.

The sun was in the west when this began. Now it's almost setting. Again we hear the distant call, faint and wavering. Someone in the distance wants to be heard, but no one answers. The sun moves slowly below the horizon. Whoever was calling must be discouraged. The voice in the distance stops.

Soon it's dark. The waning moon will rise but isn't up yet, so only stars are showing. But they're in the wrong places! Where is Orion? Something has happened to the sky. I hear the heavy, continuous rustle of bushes shaking and branches cracking. The sound gets fainter. The dinosaurs are moving away.

Now I'm sitting in my house but I don't turn on the lamp. I remember my cousin Tom once saying that the stars are moving, and over the millennia have moved considerably. I'm seeing them now in their modern places, and their light I saw through the trees by the river is starting to come through my window.

CHAPTER 15

Pterosaurs

Perhaps the most fascinating products of the archosaur line were the *Pterosaurs*.* Some of these were huge, and all could fly. They lived at the time of the dinosaurs and at first were thought to be dinosaurs. But as it turned out, they were not. They were a unique form of reptile, and even more interesting than dinosaurs, if that's possible.

One wouldn't realize that though, if one were judging from what was said about the first known pterosaurs. These were known as *Pterodactyls*.** When I was in grade school and learning about dinosaurs, according to some popular books, pterodactyls were dinosaurs who flew. But not only were they thought to be dinosaurs, at least by our teacher; they were thought to be clumsy, incompetent dinosaurs. Or such was their public image.

As children, we were told that their legs were so weak they couldn't take off as birds do by jumping into the air. Pterodactyls had to roost on cliffs, we were told, where they hung by their feet like bats. When they wanted to fly, they'd let go with their feet and drop through the

---

* *TERR-oh-sores*—"winged lizards."
** *Terr-oh-DACK-tills*—"wing fingers."

air to get started. I still have a mental image of a pterodactyl as a mid-sized creature with a beak, hanging by his feet from a cliff, deciding to fly, letting go, falling down the face of the cliff as he spreads his leather wings, then flapping clumsily before collapsing on the ground where he pushes himself along on his belly, hoping no predator sees him. The helpless pterodactyl seemed less competent even than the overweight brontosaur in the pond. That both these creatures went extinct seemed inevitable.

Over the years, it has struck me as strange that paleontologists would conceive of such helpless animals, so I now feel sure that my childhood concepts didn't come from scientific theory. But the part about hanging from a cliff did, perhaps because pterodactyls were discovered in 1784, before we had airplanes, and we couldn't imagine something like a pterodactyl just taking off and flying. So I was glad to learn my image was a misconception. One thought one wouldn't like a pterodactyl any more than one would like a brontosaur, when in fact the opposite is true.

More was discovered—much more. Pterodactyls belong to the *Pterosauria* group, which produced, without a doubt, some of the most fascinating, most capable animals ever to grace our planet. They left no evolutionary heirs like birds or flying lizards, so nothing like them still exists today. Their kind appeared soon after the P-T or Permian-Triassic extinction (the "Great Dying") and disappeared during the next great extinction, the K-T or Cretaceous-Tertiary (the fireball) extinction. Their fossils are found throughout the world, and their disappearance must be seen as one of the world's great tragedies, not only for them but also for us, as we will never know them.

Perhaps the most impressive report about pterosaurs is by paleobiologist Mark T. Witton.[*] At the time of this writing he is based at the University of Portsmouth in England, and his book, *Pterosaurs: Natural History, Evolution, Anatomy*, is riveting. Shining with responsible, carefully

---

[*] Mark Witton, *Pterosaurs: Natural History, Evolution, Anatomy* (Princeton: Princeton University Press, 2013).

documented science (although of course, as normally happens to innovative work, it has a few detractors with different and perhaps outdated opinions) it shows the underlying confidence of someone who knows what he's doing and is going to do it anyway when he calls a pterosaur's front feet "hands." Why risk public criticism for this anthropomorphic folly? Because they *were* hands, and he's happy to say so. Not for him the sewer of anthropodenial, which insists that an animal has no hands, just front paws, because a pterosaur's hands were nothing like his feet. Instead they appeared at the bend of the wing, which was stretched by a very long finger.

"The pterosaur forelimb contains the same bones that you have in your own arms, although they are so modified for flight that you could be forgiven for not recognizing them,"* says Witton. So I grind my teeth in frustration, wishing pterosaurs had been accessible sooner. I think how, as a kid enthralled by dinosaurs, I would have loved to learn about them. My brother and I would have imagined ourselves riding on the back of a big one and going to South America. But they had yet to be revealed to all but a few, and what a revelation it was! They were as different from us as it's possible to be but at the same time were very much like us! I look at my forelimb. It's not quite like a pterosaur's, but it has the same bones! I rejoice that pterosaurs and I are related—not closely, alas, but related! With a little imagination, I see us in the same clade, however vast, because we share a common ancestor, an unknown tetrapod mother who laid an amniotic egg that hatched three hundred million years ago and aimed her lineage at pterosaurs and me.

Perhaps the single most impressive product of evolution the world has yet to see was not the human but the pterosaur. The early pterosaurs were small, and not all were adept fliers, but they eventually evolved an enormous winged creature, the azhdarchid.** An azhdarchid was about the same height as a giraffe and was to some extent built along

---

* Witton, 32

** *Az-DAR-kid*—said to mean "dragon," but could also mean "from Azhdar," a village in Iran.

the same lines with a very long neck, very long legs, and a relatively thin, strong body.

A famous drawing* has made its way to the internet, a drawing that shows Witton himself standing to the right of such a pterosaur, the scientist's head not much higher than the pterosaur's low-level elbow. To the pterosaur's left stands a huge Massai giraffe, whose head is right beside the pterosaur's. Both are at least fifteen feet tall, and with their long legs and necks they look rather similar. But the pterosaur's beak-like jaws are at least half as long as the giraffe's neck, and the giraffe is more massive. The pterosaur's arms have wings attached, but in the drawing, he's in a quadrupedal stance, his wings are folded, and his hands are on the ground in front of him.

Pterosaurs had long, straight arms and legs, although in the wonderful drawing his arms and legs are slightly bent. And his wings! If he spread them, they would span about thirty-four feet.** Evidently these were made of muscle tissue, not skin tissue, so they weren't like leather as was first supposed, and they stretched from his arms to his thighs. If we had wings like that, they'd stretch from the undersides of our arms to the sides of our shins and would fold to stick up along our backs when we were standing on all fours.

Flight is a great convenience—so useful, in fact, that one wonders why more of us can't fly. Why didn't we evolve from flying squirrels, for instance? We, too, were small when we lived in the trees, leaping from branch to branch as squirrels do. It was even suggested that one of our ancestors was like a flying squirrel, with skin from her arms to her thighs. So if evolution was even halfway fair, we too could have developed wings and be flying.

Don't say we're too heavy. Some flying pterosaurs were five times as heavy as a present-day human. We wouldn't have to sit for hours in the airport while the plane was delayed—we could just run a few steps and take off. And then, rather than spending most of a day

---

* Witton, *Pterosaurs*, 250.
** Witton, 251.

slipping and sliding down a steep, rocky canyon and struggling up the far side, we'd whisk right over it in seconds. We'd float through the sky over rushing rivers that would sweep us downstream and over a waterfall if we were brave enough to try to swim across. And with the wind behind us, we'd soar above mountains as big as the Alps or the Andes or even the Himalayas, instead of spending tremendous energy climbing only as far as the first bivouac—to say nothing of the energy we'd spend over the following days and weeks, always in danger of avalanches or crevasses, risking our lives as we clawed our way up and over the massif until we were on flat land again with the mountain range behind us.

To know what this would be like, we turn to the horrifying accounts of Himalayan first ascents such as those of K2 and Annapurna. We would have climbed over the sides of such mountains, not up to the summits, but even so, exhausted and starved, with our hands and feet frozen and gangrenous, we could then look with envy up at the sky and think of pterosaurs. They easily flew above such mountains, no doubt enjoying the updrafts. "Geographical barriers would mean nothing to these guys,"* says Witton.

According to some sophisticated scientific work by Witton and Michael Habib, a specialist in biomechanics, azhdarchids could fly at fifteen hundred feet, traveling for perhaps ten thousand miles at seventy miles an hour, and staying in the sky for five or six days at a time without landing to rest or eat. This was quite an achievement. No gasoline-powered plane can fly for five days without refueling. In 2015, solar planes began to set records, but they were fueling from the sun. Yet millions of years ago, the pterosaurs did even better, "refueling" before takeoff, presumably with normal foods. Considering that a big pterosaur could weigh five hundred pounds, they were far and away the biggest life-forms who flew at any time in any form ever.

But why did they travel? Surely for the same reasons that modern birds and mammals travel, to where they found seasonal food. Some

---

* Witton, 256. Witton is paraphrasing Mike Habib; see below.

pterosaurs ate fish, for which in winter they'd need to fly to ice-free water; others ate lizards, which would be everywhere in a warmer climate but out of sight in winter. Today, many kinds of animals migrate seasonally, avoiding unwelcome climate conditions and taking advantage of available water or seasonal foods. Dinosaurs did this too. Pterosaurs might have starved if they hadn't done this, and the big ones would have been the winners, as they could travel very far, even from one hemisphere to another.

Food and climate surely influenced the pterosaurs, who seem to have been self-warming, judging by the fact that their bodies were coated with a kind of fuzz—not exactly feathers or fur, but what in science-speak is known as "pycnofibers."[*] It wasn't exactly fur, but evidently it looked like fur, and it told the scientists something important: not only does fuzz suggest self-warming, but it could have been a consequence of flight. The energy needed for flight requires high metabolism, which in turn causes warmth, which in turn requires some kind of cover on the skin to contain it, given the relatively thin skins we inherited from the chain of our ancestors starting with fish. Or as Witton more mindfully puts it, "The evolution of powered flight in birds and bats seems to have followed prior development of active lifestyles and elevated metabolisms, so the same may be true of pterosaurs."[**] Thus pterosaurs seem to have been warm-blooded.

The pterosaurs' flight inspires us of course, but these animals also did plenty of walking. While some pterosaurs are thought to have caught fish by diving after them, others seem to have hunted on foot and thus have been likened to storks, not only because storks hunt on foot, but also because some pterosaurs looked like storks. This was especially true of azhdarchids, who had huge, thin jaws that looked like beaks. Their jaws, however, were about eight feet long. Aside from that, they looked something like storks.

---

[*] *PICK-no-fibers*—"dense fibers."
[**] Witton, *Pterosaurs*, 19.

Witton tells of fossilized tracks that a pterosaur left in soft mud.[*] This pterosaur flew down and landed on his feet. Because, no doubt, he still had momentum, he then took a hop and put his hands down as if for balance. He must have folded his wings, because after that, with his hands on the ground in walking mode, he strode off on all fours, taking regular, confident steps as if walking was as natural to him as it is, say, to us.

A pterosaur's hands were at the bend of his wings. His hands had four fingers but his tracks show only three. So the fourth—the extra-long, strong finger that extended his wing—must have been up along his side to fold his wing, giving us another glimpse of what he'd been doing.

How exciting it must be to find such tracks and interpret them, to gain insight into several seconds of a pterosaur's life! Such tracks are the image of a moment in time like a photograph, or even like a film because they capture motion.

Other sets of tracks show that pterosaurs walked in the way of many mammals—foxes, for instance—because the print of the hind foot can appear ahead of the front foot. If fox-prints look like that, it means the fox was in a hurry. If he's not in a hurry, the prints of his hind feet fit partly over the prints of his front feet. A fox in a hurry must run, and of course a pterosaur in a hurry could fly. Still, although it's somewhat speculative, it's easy to imagine that pterosaurs walked and even ran quickly, in which case, their speed would come as no surprise. Because they ate mid-sized reptiles, the pterosaurs had to probe around in bushes, looking here and there for motion. Like the lizards of today, many reptiles of the time were mid-sized and probably very fast, and a pterosaur would have been fast and agile too, if he caught them.

So marvelous a creature fires the imagination, and has inspired a number of people to claim they have seen one. According to them, certain pterosaurs did not go extinct but are hiding in a remote part of the world where scientists can't find them. With them, no doubt, are

---

[*] Witton, 63.

Abominable Snowmen, Loch Ness monsters, American Big Foots, and extra-terrestrials from outer space who came in flying saucers.

If only it were true about the pterosaurs, though. It seems very human to believe it's true. No one, as far as I know, claims to have seen a prehistoric spider or a prehistoric mouse. It's very human to select a large, exciting creature to believe in.

CHAPTER 16

Crocodiles

We've been thinking of the time between the Carboniferous and the Cretaceous, a period of about 150 million years, in which lived our proto-mammal ancestors—the wide assembly of different-sized synapsids with and without sails—as well as dinosaurs and pterosaurs. The only large animals not mentioned so far that belong to this era are the crocodile-types—a group which today includes alligators, gharials, and caimans.

Although descending directly from the early archosaurs and changing somewhat as time went by, crocodiles have not changed nearly as much as most other kinds of animals. They were crocodiles then and they're crocodiles now, so maybe (as with the millipedes) they were successful to start with and stayed successful all along.

Very early on, crocodiles sprang from their archosaur ancestor as a separate group from dinosaurs. At first they were fairly small creatures, many different kinds of them. Most if not all were terrestrial, and some seemed to be halfway between crocodiles and lizards. Some had sails, some walked on four legs, others walked on two legs like some dinosaurs, some ate animals, others ate plants, and some even moved toward self-warming. A possible ancestor of modern crocodiles, *Xilousuchus**—an archosaur who lived about 250 million years ago—was a small carnivore with a sail who walked on four straight legs, weighed about five pounds, lived on land, and seems to have been self-warming.

But his big, predatory descendants—the crocodiles we know today (if in fact they are his descendants)—got their start toward modernity about two hundred million years ago, when Pangaea was breaking up slowly. With more coastlines developing and infiltrating the land with water, the air masses didn't lose their moisture while moving inland. So the climate was changing from dry to reasonably humid. Many life-forms were reconsidering their evolutionary paths, and the successfully expanding populations of the dinosaurs may have moved the crocodile-types back to the water.

Some of them grew to large sizes, gave up on self-warming (probably because it's nutritionally expensive), and even gave up on straight legs, although to this day they still straighten their legs. However, they seldom do this except to move quickly—then they "high-walk," they even gallop—or to look over an obstacle, as we would by standing on tiptoe.

Legs that are normally bent help crocodiles in several ways. You can run a bit faster, your stance is more stable, and it's easier to climb up a riverbank than it is with straight legs. Bent legs also help when in the water. Like the eyes of many animals who spend time on land and water—frogs, mudskippers, and even hippos—a crocodile's eyes are near the top of his head, so here again, bent legs are beneficial. If standing in shallow water, a crocodile can straighten his legs for a quick look around and then relax. That way, he can learn what's going on while remaining inconspicuous. While he's relaxed, just his eyes and maybe the top of his head are above the water. And his eyes don't really look like eyes. They look like nothing to worry about, as if they were floating debris.

---

* *ZILL-oh-SOO-koos*—"crocodile-type found near Xilou River."

A crocodile can trot on straight legs, sort of, but if he's really in a hurry—if, for instance, he's sunning himself on a riverbank, gets scared, and needs to get back in the river for cover—he slides on his belly, knees bent, elbows out, scrabbling with his arms and legs and lashing with his tail for extra propulsion. In the blink of an eye there's a splash and he's gone. Why doesn't he stand up and run? Perhaps because the banks are often slippery. You can fall if you try to run down them. It's safer and faster to slide.

Why a small, straight-legged terrestrial animal would evolve to become a huge, elbows-out aquatic animal might seem unclear, but crocodiles were not the only ones to do it. Otters, platypuses, whales, and dolphins made the same kind of decision, so it must have appeal. The result has been that to this day the crocodile-types are more like their early ancestors, the issue of those long-ago archosaurs, than they are like the other archosaur descendants. And in many ways, some of the early crocodile ancestors looked very much like modern crocodiles despite their size and habits. Crocodile-types were on Pangaea 250 million years ago, and now they're on the continents that broke away. Few modern animals can say the same—it's a very long time of survival. But then, if you're an enormous animal who can live for months without eating and can hide in a river which edible animals must visit to drink, not realizing that you're watching, you're an excellent candidate for survival.

But crocodiles may not survive us. We're destroying their environments and otherwise exterminating them. We hate and fear them so we shoot them when we see them, and we want their skins for fancy shoes and purses to go with our mink coats. But if not for us and our limited priorities—we don't care who goes extinct as long as we're in fashion—crocodiles would be named among the world's most successful animals, right at the top of that league. They'd be way ahead of us in that respect—if they had time to evolve unfashionable skins.

We humans had nothing to do with crocodile evolution. But it's we who sometimes benefit from the way their legs evolved, as this author and her daughter (who is paraplegic and uses a wheelchair) were to learn while camping by the East Alligator River in northern Australia. At the time, we didn't know the name of the river so we put our sleeping bags on the ground.

During the night, we heard an awesome gush of water, like a big wave breaking on the shore. Anything making a wave like that cannot be good, especially not at night, but we didn't know then what we know now—that the river is home to the world's largest land-based predators, twenty feet long or longer, weighing over two thousand pounds, and responsible for dozens of human deaths annually—the Indo-Pacific crocodiles.

One had come out of the river. I don't think he knew we were there—he was taking slow, thoughtful steps—but he seemed to be walking toward us. He must have been walking with his knees and elbows bent, because thanks to my pillar-type hind-limbs and the strength of my daughter's long, straight forelimbs that propelled her speeding wheelchair, we moved very fast and got back in the car. For this we thank an ancestor not shared by that crocodile, an ancestor with flexible limbs that were strong and straight, who didn't walk with her knees bent and her elbows out, and who didn't get caught by a crocodile, or not before she'd passed along her genetic information. And a good thing, too. For a while we heard something big—perhaps his tail—dragging slowly over gravel near the car.

CHAPTER 17

Birds

Sixty-six million years ago, our world experienced the K-T or Cretaceous-Tertiary extinction. Who knows what our planet would have been like without it? A favored theory holds that the extinction period was brief but world-wide and thorough. Seventy-five percent of all life-forms disappeared, including plants, land animals, and sea creatures—or just about everything except insects.

The catastrophe seems to have been caused by a space-rock six to nine miles wide striking the earth near what is now the Gulf of Mexico, leaving a crater more than a hundred miles wide. One theory holds that the strike blew up so much land that hundreds of square miles of matter flashed up above the stratosphere. One theory holds that it then caught fire as it came speeding back down through the atmosphere, making parts of the Americas suddenly go up in flames. It also caused tsunamis and released climate-altering gasses that flew out of the rocks that were obliterated, leading to acid rain and global cooling.

Whether or not some of the matter came down in flames, much of it would have shot up as dust and stayed there, causing years of global winter and the world-wide death of plants, who couldn't get enough sunlight. And then, because plants are at the base of the food chain, the loss of the food that plants provided would have killed most of those who survived the fire, if indeed there was a fire.

As for years of global winter, plants with seeds might have done fairly well, because while a parent plant may die, the seeds it has distributed can often live a long time and don't need favorable conditions to do so, especially if they're buried. Witness the seed of *Silene stenophylla*, the little flower found in the Arctic, mentioned earlier. It sprouted after thirteen thousand years.

As for the animal survivors, seeds probably contributed to their fate in several ways, not only by sprouting after things cleared up, but by providing food for some until that happened. Little creatures don't need much food, so small herbivores could have lived mostly on seeds and in turn could have served as food for small carnivores. Some little animals could have stored food as squirrels do, although how long such a cache would last is debatable. Animals such as snails, worms, and some insects could have survived by eating decaying matter, and then could have been food for slightly larger animals. Thus various kinds of animals made it through.

Crocodiles made it, despite their size. Not only do they live in water, but they also keep the ancient reptile custom of burying their eggs. Their young, when hatched, are reasonably self-sufficient. Mother crocodiles protect their hatchlings, but if important predators were no longer present, the hatchlings would have needed less protection. Then too, crocodiles can go for months without food—a huge advantage not available to the warm-blooded—and are also glad to eat carrion. This would have helped, of course, if some of the other animals died out slowly, providing a steady supply of carrion over time.

The pterosaurs, the large- to mid-sized dinosaurs, and even many smaller dinosaurs including some bird-like ones, did not live underground or in water. They needed food in the form of healthy, growing plants or mid-sized animals, and it was they who vanished forever.

By the time this extinction occurred, several groups of birds were already present and many survived. Interestingly, one doesn't find answers to the question of birds surviving a fireball extinction, although some of them did. It's been suggested that back in those days, all birds buried their eggs, but this is not confirmed, especially since the bird-like dinosaurs are thought to have incubated their eggs. Birds of the chicken and turkey family bury their eggs, but it's been suggested that they recently began this, and may not have done so at the time of the disaster.

The first known non-pterosaur flying vertebrate was the famous *Archaeopteryx*,* an animal with long jaws, teeth, and a long, boney tail, much like his dinosaur forebears. The *Archaeopteryx* appeared about 150 million years ago, long before the fireball extinction, during the fruitful Jurassic. They are not considered to be the ancestors of birds that followed, although they are classified with birds. However, more birds were to develop, and twenty million years later, during the mid-Cretaceous, they had evolved to fill many different environmental niches. In other words, long before the fireball extinction, their species and numbers had expanded.

Some Cretaceous birds looked something like *Archaeopteryx*, but others were beginning to look like modern birds with no teeth, strong shoulders for flight, and hollow bones. They got the hollow bones from dinosaurs who may have used them for oxygenating purposes as birds do today. But the hollow bones they inherited from their dinosaur ancestors might also have helped with weight reduction in birds, who came to have short tails for that purpose which, although boneless, grew extra-large tail feathers, not only for support while in flight, but also for balance while walking.

Birds had moved in many directions, to the shores to catch fish, to the savannahs and forests to find insects or seeds, and to the skies as large predators along the lines of modern hawks and eagles.

But then, as has been true of many taxonomic groups in which some members have reversed themselves—like crocodiles, say, if they gave up pillar-type legs and self-warming—some birds gave up flight.

---

* *ARR-kee-OP-turr-ix*—"ancient wing."

Today, the trend is seen in large flightless birds such as ostriches, and was formerly seen in others such as the now-extinct but highly successful terror birds. These appeared after the fireball extinction but were very much like some non-avian dinosaurs, as if the mid-sized non-avian dinosaur model found success at different times. If the first such creatures vanished with the fireball, others like them evolved to take their place. When Gaia has a good idea that fails because of an extinction, she often tries again.

One wonders if certain dinosaurs were replicated by terror birds. A life-form expanding often turns to large sizes, and terror birds were ten feet tall and evidently could run at forty miles an hour. The fastest human runner on record ran at twenty-eight miles an hour. But most members of our species are slower, and after we arrived in South America, we coexisted with them. They were bigger and faster than we were. Did they prey upon us?

They certainly could have caught us. With their heavy beaks and sharp claws, they are thought to have killed their prey by running it down, knocking it to the ground, killing it by powerful kicking and hammer-like pecking, then pulling it apart with their claws. These birds became apex predators, certainly in South America where most of their fossils are found.

Their decline began after a volcano blew up enough lava to make the Isthmus of Panama. Across the isthmus came carnivorous mammals from North America who then, along with the resident alligators, became serious competitors for the terror birds' prey. It's also said that the arrival of our species in the Americas contributed to their extinction. They might have killed us, but we must have killed them when we could, especially if we knew that if we didn't kill them they could knock us to the ground, kill us by hammer-pecking, and pull us apart with their claws.

Today, if you watch your birdfeeder with all the little goldfinches sitting on the perches eating seeds, you seldom think of dinosaurs.

Goldfinches don't fit the image. Therefore the connection, at least to carnivorous dinosaurs, may best be seen in birds of prey. To my mind, there's something about a falcon that suggests a dinosaur. A falcon has a kind of constant, single-minded focus and determination that you seldom notice in a mammal. So if you want to see up close what some predatory dinosaurs might have been like, you could find a suggestion in falconry. This, while somewhat murderous, is nevertheless exciting.

To encounter such a bird, you visit a falconer who will lend you a glove to wear on your left hand and arm. You will hold out your gloved arm and a flying predator, an avian dinosaur, will land on it. But first you must carefully prepare. Your gloved hand holds some kind of bait such as a dead baby chicken. The hawk is sitting in a tree while you fish in your pocket to find the chicken—no doubt she knows you're preparing for her—and you then turn your back and extend your gloved arm. She will be mildly excited. She will fly down and land on it. But you must watch her approach in case she does something unexpected. You must turn your face away and watch from the corners of your eyes because your eyes should be as far from her oncoming claws as possible.

The well-known New Hampshire falconer, Nancy Cowan, said why. "That's how many hawks kill their prey," she explained, "forcing their talons through the skull via the eye sockets into the brain. If they think there's a chance the food might go away, they fly in your face and go for your eyes."*

A statement like that gets one's attention. An eagle in Mongolia killed his owner's grandson by this method, in revenge for which the owner cut off the eagle's feet and released him, and fossil evidence shows that raptors used this method to kill the children of some of our near-human ancestors.

Even so, I've had the extreme pleasure of visiting Nancy Cowan and flying her Harris's hawk, Fire. Fire was high in a tree, standing on a branch and watching. I made sure to follow Nancy's instructions,

---

* Sy Montgomery, *Birdology* (New York: Free Press, 2010), 129.

turning my back so my face wouldn't get involved, taking care to look over my shoulder, spotting the hawk with my peripheral vision, and only then holding out my gloved arm.

On motionless wings, Fire slid through the sky. Gosh, but she was big—her wingspan was longer than my outstretched arm. Suddenly she tipped up to throw her legs forward and slammed her feet on the glove. Her claws gripped hard. Her grasp was tight. She was strong. She was heavy. And I was thrilled because a big, dark hawk was standing on my arm.

She took the chick. I turned to her. She gulped the chick, then settled her feathers and looked at my face. But her gaze was straight on. Her pupils didn't move as ours might when we examine something. This seemed unusual, but birds' eyes are better than ours, and birds of prey have the best eyesight of all. Since Fire had large eyeballs while mine were small, her visual field was wider. She didn't need to twitch her eyes—she was getting the picture by keeping them open.

But could we communicate? My face must have showed how excitedly happy I was, but a bird might not realize that because they don't have facial muscles. In birds, the little tell-all movements of a mammal's face are missing. As Fire looked at me while I looked at her—the cross-gaze of a mammal and a dinosaur—I was somewhat surprised to find that she was just staring. Maybe she could read my face and imagine she was communicating—assuming she saw value in communicating with something such as me—but I imagined no such thing because I couldn't read hers.

Mammals don't like mutual staring. They take it as disapproval or a threat. Fire wasn't a mammal, but I am, so I shifted my gaze. Even so, I was fascinated with this hawk and wanted to repeat what just happened, so again I raised my arm. She spread her wings and flew to the branch of another tree.

She would have acted this way in the wild, probably because she'd have few other options. If nobody with a glove was waiting, she'd fly to a branch, and if she saw nothing of interest she'd take to the sky and soar. She'd be looking around, of course, and if a mile away a squirrel was keeping still among some leaves, she'd see him. The eyes of a hawk are bigger than her brain, and in some ways, more important.

I reached in my pocket for another dead chick. When I raised my arm again, Fire flew back and landed on it. For her, the gulps of meat were her big moment.

But my big moment was seeing her eyes. Even thinking about those eyes makes my skin prickle. This means my body wants me to look bigger, but why? I had no need to seem daunting, and she had no objection to standing on my arm.

Even so, deep in my heart was a tiny flame, not quite of fear, more like a small, ever-burning votive candle. I was not in danger, but the tiny candle flickered just the same. Was I conjuring my ancestral past when raptors killed the smallest of us? Or was it her face, her motionless face, as still as the face of a statue? For a very long moment she seemed like a statue, but no, she was a living, breathing hawk with a big, yellow beak, two little nostrils, and large, round dinosaur eyes that could see for miles and miss nothing. But how did she feel? Did she like me? Would she know me if she saw me again?

"Please, Gaia," I wanted to say, "I'm just a mammal. I can only learn by watching other faces. But here's this hawk whose face tells me nothing. How can I understand her?"

And Gaia said, "I made that face a long time ago, when nothing like you was anywhere. Why would I care if you don't understand her?"

So I raised my arm and Fire flew to a tree. I turned my back, found another dead chick, and raised my arm again. Then a big, dark hawk sank her claws in my glove, gulping a chick who lost her life for this occasion. But the hawk was a hawk, and though I felt the little flame, I was deeply excited. That wasn't enough? Did I need to understand her?

CHAPTER 18

Mammals

The word "mammal" comes from "mamma," the Latin word for breast. Today, various forms of "ma" and "mamma" show up in various languages as a synonym for "mother" because "ma" is often the first word a baby says. After all, if you open your mouth while making a noise, "ma" is what comes out. When a baby does this, we think she's calling for her mother. The Romans must have thought she was calling for some milk.

The Roman translation of the baby's remark prevailed, at least in English, and today the word "mammal" refers to milk-producing glands inside the breasts of those who have them. But in animals like platypuses, who don't have breasts, these glands are patches under the skin—a gift from our ancestral synapsids. And if at first these glands leaked moisture that might have been ancestral sweat, by the time the modern platypuses got them, they leaked milk.

Long before the fireball extinction, those of us with milk glands (or in other words, we mammals) were developing in different directions. One early group, now considered the most primitive, includes platypuses and echidnas, and before going further, I should explain what these animals are like.

I once saw a platypus in Australia, or I think I did. I wish I could describe some great adventure, but all I saw for just a few seconds was a streamlined little someone slide by in a narrow brook that ran under heavy vegetation. *What's that?* I wondered, not knowing I may have been looking at an early type of mammal, an echo of the creature who was far from the fireball when it hit, sleeping in her burrow in the bank of a stream.

She would have been upset while everything was burning, and for days she would have hidden in her burrow. But since platypuses eat crabs which are highly nutritious, she could wait a long time before eating. In time she would cautiously emerge to hunt for crabs because they also could have survived the fireball. They hid in wet leaves when they weren't in the water, and in this scenario, they lost their lives so a platypus-type could return to her burrow and hide.

The platypus line gave rise to the echidnas, and echidnas have pouches like kangaroos and other marsupials. There are now five species of these round little creatures, and all live in bushy environments. The Australian echidnas have spines in their fur and look something like prickly soccer balls. Below their little eyes, they have long, thin snouts for pushing into termite mounds. They eat the termites, also ants, worms, and various kinds of insect larvae.

Both platypuses and echidnas mate once a year and lay eggs once a year. Normally a platypus lays two eggs and an echidna lays one, which she carries in her pouch. The platypus has a flexible tail with which she presses her eggs against her hindquarters. The eggs hatch fairly quickly, and because the crabs eaten by the mother are so nourishing, she won't starve while she waits.

Neither platypuses nor echidnas have breasts, but they do have milk glands, which qualify them as mammals. The platypus's glands leak milk into wrinkles on her belly and the infants lick it up from there. The echidna's glands are in her pouch, ready for the infant to suck when he hatches.

Echidnas and platypuses are known as monotremes, a word from Greek which means "single hole" and refers to the cloaca. Thus they're unlike the rest of the mammals and more like birds or reptiles, with one exit for turds, urine, and offsprings, not two separate exits like we have.

Fossil evidence is sketchy, but about a million years after the monotremes appeared, someone from the same lineage (if not the same branch) gave rise to marsupials*—kangaroos, koala bears, Tasmanian devils, and many others—and also to placental mammals such as us. What makes our type of mammal different from the monotremes? As many of us know from personal experience, when we mammals go into labor, an infant emerges instead of an egg.

We placental mammals have a different strategy than the marsupials, however. Our placenta, among other advantages, allows our infants to stay in the womb longer than marsupial infants do. An unborn placental mammal is fed by its mother's blood supply, almost as if it was part of her body. An unborn marsupial is fed by a sort of egg-yolk arrangement which gets used up in time, and perhaps for this reason, a newborn marsupial looks to us as if he's still an undeveloped fetus.

Throughout this book I've been dismissing the concept that animals are pre-programmed, and now I'm sorry. The best example of what seems to be pre-programming that I know anything about is a baby kangaroo, who seems to know (there's doubt here) when he should emerge from his mother and almost seems to do so on his own. Perhaps his egg-yolk food supply ran out and he got hungry. Or perhaps his mother squeezed him a little with contractions. But the moment he's born he's ready to travel, so to speak, so even though he looks like a premature fetus, he climbs upward through his mother's fur until he finds her pouch. He goes inside and there he finds her nipples. She has four. He'll move from one to the next as he matures.

When he's a little older, he will look out at the world from the top of her pouch, and later he will climb out, to return when he's doubtful or hungry. Does he learn about the world as they move around together, discovering what to eat and what to avoid, what to fear and what to

---

* From Latin—"pouch" or "resembling a pouch."

challenge? By then, he's way past whatever he was pre-programmed for, so surely by the time he leaves his mother, he's ready for the world.

Marsupial fossils are found world-wide but with different histories and different distributions due to the various post-Pangaea ecosystems. In certain ecosystems the placental mammals are thought to have outcompeted the marsupials where the two kinds of mammals coincided. The Americas, for instance, have only one kind of marsupial—the possum, if with several species—while Australia has dozens of kinds of marsupials with hundreds of species.

It's said that marsupials do better than placental mammals in dry climates, and most of Australia was reliably dry. In fact, except for bats, monotremes and marsupials were the only mammals in Australia until people began arriving, bringing more placental mammals such as dingoes, rabbits, deer, foxes, dogs, and cats. Here, we humans played the role of, say, a volcano or a glacier, causing a bridge between two landmasses, giving the animals on one side access to the other side, where newcomers would compete with the residents with varying degrees of success. The imported placental mammals, especially those such as rabbits and feral cats, pose serious problems for Australian marsupials today. And dingoes are thought to have out-competed the Tasmanian wolves, contributing to their extinction. Something like this had happened before. After volcanic action created the Isthmus of Panama, placental mammals from North America went to South America and are presumed to have outcompeted the South American marsupials until only the possums remained.

Today, there are more than a hundred and fifty kinds of marsupials. They're found in Australia, Tasmania, and Papua New Guinea (as well as in the Americas), and they appear in all forms and sizes. The long-tailed planigale from Northern Australia, a marsupial, is the second smallest mammal ever found. He's less than two inches long, looks like a mouse with a flattened head, sits up like a squirrel, and hides by day in grassy tangles only to leave at night to prey upon insects and their larvae. (The smallest mammal ever found is placental—the bumblebee bat of Thailand, who is the size of a bumblebee and weighs less than half an ounce.)

The biggest marsupials today are kangaroos. The favorite marsupials are probably koala bears, because they're adorable, and Tasmanian devils, who are said to make excellent pets unless you annoy them. A dangerous marsupial of the distant past would have been *Thylacoleo carnifex*,* a fearsome cat-type carnivore who weighed three hundred pounds and had dagger-like teeth, front legs that were very much like human arms, and fingers with retractable claws. They seem to have coexisted with humans, however—an Aboriginal rock painting shows a lion-like creature with claws and a tufted tail.** These remarkable animals are thought to have gone extinct maybe forty thousand years ago. But some people claim to have seen them, perhaps in the land of the pterosaurs, Big Foots, and Loch Ness monsters.

But no wonder this marsupial was fun to imagine. A big, ferocious lion-type must have been something to see, especially if a little, ferocious lion-type was peeking out of her pouch. And imagine those retractile claws! Of the millions of animals now on the planet, only cats and a few kinds of weasels have retractile claws. This means that somewhere, a marsupial must have evolved such claws on its own. How's that for convergent evolution?

Without going to Australia, the only marsupials North Americans can ever know are possums. Placental mammals may have outcompeted most of the marsupials in places where they occurred together, but no one has outcompeted possums or even preyed upon them with much success.

Possums can't run fast. If they're not safely in a tree when confronted with something dangerous, they play dead and seem for all the world like corpses. Their lips are slightly retracted and their eyelids are not quite shut, so they look like the lips and eyelids of the dead. Foam

---

* THIGH-lakka-LEE-oh CAR-nif-ex—"pouched lion executioner."
** Peter Murray and George Chaloupka, *The Dreamtime Animals: Extinct Megafauna in Arnhem Land Rock Art* (Sydney: Oceania Publications, 2017).

leaks from between their teeth and a rotting smell leaks from their anal glands. If you pick them up, they just dangle. The condition is said to be like fainting, but it looks much worse and is very convincing. It works often enough with animal predators, but not with oncoming cars, today the main predators of possums.

On a roadside in Virginia, my daughter and I once found the body of a mother possum killed by a car. Seven infants that looked like fetuses were clinging to her fur. Five were still living, so we took them home, made a pouch from a cotton sock inside a woolen sock, and put them in, removing them hour by hour, night and day, to feed them artificial cat's milk with an eyedropper. In this way, we raised them from naked little fetuses to be furry little juveniles that looked like small adults.

We intended to release them, so we tried to teach them about the outdoors. But we hadn't the faintest notion of how their mother would have taught them except by example, and that we couldn't provide. They did learn to hide in a woodpile if they felt threatened, so we thought they'd be okay if we left them for a little while when we went to do some errands. But try as we might, we still hadn't been able to teach them how to climb a tree, and one day a dog found them and killed them, leaving their uneaten corpses where they lay. Maybe they tried playing possum while he did this, and if so, he killed them anyway. Or else they must have watched him helplessly, not knowing they could save themselves by climbing a tree.

We didn't see this happen, and the dog wasn't really to blame—he wasn't our dog and didn't know the little possums—he just did what dogs do. I was to blame for leaving them alone, and I've been wracked with guilt since then. If I could undo the past and relive a moment, I'd go back to that.

As for us placental mammals, although there were plenty of fossils that seemed closely related to our placental ancestor, an international team of scientists made a complex study of genetics, molecular data, and physical characteristics to replicate the actual ancestor, and came up with a rat-sized model with a long tail who lived in the trees eating insects—something that looked like and probably acted like a shrew. Quite recently, a fossil of the real thing was discovered in China, dating from 125 million years ago. It looked like a shrew and had the kind of teeth that suggested an insect diet, so the replica was successful.

As the continents moved, the early placentals diverged. They were small at first, but many became bigger, and today more than four thousand species exist, from the world's smallest mammal—the half-ounce bumblebee bat from Thailand—to the two-hundred-ton blue whale. Many others have gone extinct, including elephant-types, horse-types, rodent-types, lemur-types, buffalo- and cattle-types, deer- and antelope-types, sloth-types, dire wolves, saber-toothed cats, and many others, all of whom were vastly more important than our ancestral shrew was at the time.

Some shrew-types came down to the ground to become rodents and rabbits, while others stayed in the trees to become lemurs, tarsiers, and bush babies. One dubious theory holds that we evolved with or from bush babies, and although that theory is highly unlikely and has never been confirmed, I like to believe it, however hard to do, because the group is called "prosimian," which means "before monkey." And in some ways the modern ones are like the originals, as both the old and new varieties favored the forests across Central Africa. In the manner of the probable ancestor, they were and are nocturnal, with a diet of insects to which they've added bird's eggs and fruit.

So even though it's quite a stretch to say the bush baby branch is directly in our evolutionary line, I like to imagine them there because no animal is more charming. I spent half a year camped in northern Uganda where bush babies lived in nearby trees. They had big eyes, as do human babies (hence the name); they ate insects, fruits, and berries, I learned; and often enough, one of them would speak. She would be talking to another bush baby, but anyone nearby could hear what she said. That's the best thing about bush babies. They recognize about twenty different kinds of "meaningful vocalizations" (science-speak for "words"), and they usually say these one at a time or a few at a time, so they sound more like remarks than long sentences.

If one of them answered another, it almost sounded like two people talking, except that their voices weren't like ours, and the second speaker usually waited a moment before answering. So it seemed to me that a bush baby didn't just blurt out something as a human might, but carefully thought things over first. In my opinion, and because I'm so taken with bush babies, this cements their kind as our ancestors regardless of other evidence or theories.

Our ancestors are imagined as whining and grunting, but we can be sure they did better. We hate to think of anything other than a *Homo sapiens* talking, but it has now been proved that all kinds of animals, from dolphins to chickens, make sounds that have specific meanings, including sounds that refer to themselves and others, or in other words, names. Even so, human speech requires special vocal equipment that many animals lack. Some animals don't have lips, for instance. Some do (apes do) and move their lips when making a meaningful vocalization, which means that our ancestor did the same thing. But where this began is uncertain.

We were not the first to have this ability, just the first to invent so many words. English has over a million of them, although most of us use relatively few. But no matter how many words we know or use, we are by far the only animal who has carried meaningful vocalization so far, uses it so often, and relies on it so heavily. Many of us think with words, and while there's no solid evidence to inspire the assumption, it makes sense to say that animals must think with images. But who knows? Animals think all the time; but no matter how close we may be to our dog or our cat, we haven't the foggiest notion of how they do their thinking. But it certainly isn't with words.

As for the prosimians who inspired this discussion, the early ones diverged. Some became lemurs and the like, and the others became monkeys.

CHAPTER 19

From Monkeys to the Missing Link

The reader may have noticed that the author is veering toward her own kind, which she finds self-absorbed, doesn't seem to like, and can't find enough bad things to say about. But after all, she's one of them. She's also a primate, and she does respect her ancestors, especially monkeys. They seem to have begun in southern Asia and moved through Europe into Africa. From there, some crossed the ocean to the Americas, which I find to be one of the most amazing events in natural history. The author herself has crossed the Atlantic going west to east, but she was in a ship, and no one on the planet today has had an experience like that of these monkeys, so we should begin with them.

One theory holds that at a time when much of the world's water was tied up as ice, a land bridge appeared in a shallow part of the Atlantic. But others say no land bridge existed, so a favored theory holds that maybe forty million years ago, some monkeys crossed the ocean on a mass of floating debris. The debris could have formed from vegetation falling into one of the African rivers, somehow getting blocked and making a dam with water flowing around it. It could have grown larger year by year as more debris arrived, and small rodents and insects might have found homes in it. If so, they were present when unusually heavy rains turned the gently moving river into a raging torrent that swept the pile and its occupants out to sea. Several monkeys were on the pile that day, perhaps looking for berries and seeds in the fallen vegetation.

Then as now, the ocean currents moved from east to west—the right direction—and the Atlantic wasn't as big as it is today. Even so, that's a long ride. One wonders how the monkeys managed to find water. They could have found seeds and insects in the vegetation, but they couldn't drink seawater, so fortunately for them, it must have rained. Once in the New World, the voyagers gave rise to four different kinds of New World monkeys, which since then have given rise to over a hundred species.

Meanwhile in Africa, or maybe in Southeast Asia, gibbons appeared and began the ape lineage, the first monkey-types that didn't have tails. Their skulls and teeth resemble those of great apes, and their exceptionally powerful voices can literally be heard a mile away.

All but the extraordinary power of their voices passed on to the great apes, the first of whom were orangutans, who appeared in Asia about fourteen million years ago. The next to branch off were gorillas, who appeared in Africa maybe six million years later, splitting from the group that then gave rise to chimpanzees.

Chimps and bonobos are so closely related that they could almost be seen as two subspecies, just one or two million years apart, one on each side of the Congo River. Our ancestor who eased away from that group looked like a bonobo but with legs and shoulders that were something like ours. And his feet had wide heels that bore weight. This meant he stood upright quite often. But he wasn't completely a hind-leg walker—his big toes were like thumbs for grasping branches when he was climbing trees.

Why would he walk upright? He must have found some of his food on the ground, and perhaps when he was standing he looked bigger and thus more formidable than he did when down on all fours. Also, he could see farther away when standing up. But he would still have been adept at arboreal life and surely slept in the trees as protection from predators, especially the large-sized lions and other huge cats who coexisted with him.

Big cats climb trees just as little cats do, so why was this helpful? When a cat climbs a tree, just one front paw is free to hook something—the other paws are clinging to the tree. Better to hunt on the ground, charge from behind, jump on your victim, and bite through his neck.

What follows is a brief description of what happened next, perhaps rough and approximate because the science is ongoing. New finds turn up, data are revised, and fossils are re-examined, so the picture often changes. Then too, many fossils are named as if all were different species, which can be confusing if one considers modern humans who, like the fossils of old, are found in many different places and appear in different shapes and sizes. I'm guessing that if my fossil is found in what was once New Hampshire, it might not get the same name as the fossil of some Australian soccer hero found near Ayers Rock. I might be *Homo nothingmuchus* and he might be *Homo fabulopithicus*.

Interestingly, while the fossils of other ape lineages changed somewhat, they didn't change as much as ours. This seems hard to explain. Why did our ancestors keep modifying themselves for life on the ground? Bare skin exposes you to mosquitoes and sunburn, and you can't run as fast on two legs as you can on four, so you expose yourself to predators.

I know of no answer to this question, although there are theories. For instance, during glacial times, when much of the world's water was bound up as ice, rain was scarce and trees died of thirst. More space appeared between them so we couldn't swing from one to another and were forced to live on the ground. But baboons left the trees for the savannah just as we did. The climate change, if this was the reason, applied to them too, but they kept their hair and body style.

That first proto-person with thumbs on his feet may have had an evolutionary heir, by then a hind-leg walker. His fossil from 7.5 million years ago must have been found in the Republic of Chad, in the desert-like region known as the Sahel, because he was named *Sahelanthropus tchadensis*.* Other fossils much like his but from more than a million years later were found in the forested Tugen Hills of Kenya. Although during those million years, those who were fossilized must have evolved from *S. tchadensis*, their taxonomic name is new—*Orrorin tugenensis***—and their common name is Millennium Man.

All our ancestors bear the name "man" or "*homo*," as in "caveman" and "*Homo sapiens*," because in the eyes of many, everything important except housework is done by men. But "Millennium Man" is fair enough—that first fossil was a male's.

The *O. tugenensis* proto-people were about three feet tall and weighed perhaps forty pounds, which is about the size of a bonobo, but their bodies were suited for life on the ground. Their fossilized heads were set on their necks in a more upright position than the heads of those who walk on all fours. It's suggested that these proto-humans lived not only in forests but also on the wide-spread grasslands of that area, meaning that they weren't as dependent on trees as were their predecessors.

All this time, cats had been evolving, producing some formidable lion-types who were bigger than modern lions and preyed upon our kind. We would have been to them what a chipmunk might be to a Canada lynx, so it seems unlikely that most of our ancestors slept on the ground. Thus one wonders how our grassland ancestors managed.

Baboons manage. They like savannah environments, and if they can't find an area with trees, they find one with cliffs or high rocky

---

* SA-hell-AN-throw-puss chad-EN-sis—"man from Lake Chad in the Sahel region."
** Or-ROW-rin TOO-jen-EN-sis—"original man from Tugen."

mounds that a predator must climb to get them. Would our ancestors have done the same? Climbing a mound or a rock face takes time for a cat—he can't pounce when he's climbing, and he's also conspicuous. Someone in your group will see him. You can all start shouting, throwing stones, and threatening him, or you can scatter and flee in all directions. Even if he catches someone, the rest of you will escape.

Millennium Man disappeared about 5.8 million years ago, meaning that his kind survived for two million years (or 80 percent longer than our kind has been here), but during those years, they seem to have morphed as *Australopithecus*,* the "missing link" between humans and apes.

---

* *OSS-tral-oh-PITH-ik-us.*

CHAPTER 20

The Line to *Homo Sapiens*

Thanks to academic snobbery, *Australopithecus* might not have been known. The first member in this line was found in 1924 in a South African town called Taung, and an Australian gentleman named Raymond Dart recognized its importance. The fossil was that of a three-year-old child, allegedly female, although the gender may not be known. Postings on the internet confuse her with the famous Lucy who will be mentioned later, an adult female australopithecine, evidently killed by a giant cat in what today is Tanzania. The Taung child was killed by an eagle, and was named *Australopithecus afarensis*, "southern ape from Africa."

Dart found the fossilized skull in a crate containing other fossils he received from a colleague who was doing archaeological research. When Dart examined the skull, he saw that the brain would have been larger than those of other primates, and realized the skull was almost human. This was the first such fossil ever to be found, and of course he announced it to the scientific community.

At the time, he was teaching anatomy at a South African college and was also an Australian, not an established British scientist with a

PhD in paleontology. In the eyes of the British academics, he was too inconsequential to find anything as important as the "missing link" and twenty years would pass before anyone took him seriously.

But at last his finding was acknowledged, and soon after that, I met him. He was a friend of my parents, and I thought he was brilliant and exceptionally nice. Therefore I was outraged that his finding had been treated with indifference, and in doing my part to settle old scores, I can't help but repeat a vengeful analysis of the British academics who scorned him. I was told that "vanity such as theirs damages synapses in the reactive area of the brain, thus seriously slowing reaction time, and can be modified only by natural selection."

We must hope that natural selection has modified the British academics.

When at last Dart's finding was acknowledged, it was recognized as exceptionally important. For one thing, it was thought at the time that our species began in Asia. The Taung child was found near the edge of the Kalahari Desert in South Africa, where she was killed 2.8 million years ago. Other australopithecine fossils were then found, showing that they appeared perhaps four million years ago and diverged into maybe six different species (or subspecies) as they spread over the eastern side of Africa, reaching from the coastal areas almost to the continental midline, and from the Red Sea to the Cape of Good Hope.

About 3.7 million years ago, two of these hominins, thought to have been australopithecines, left tracks in what is now Tanzania. They were discovered by Mary Leakey, wife of the famous paleoanthropologist Louis Leakey. Casts of them are in Tanzania's Laetoli museum, which is near the place where those who made them walked. The actual tracks are said to be buried in a secret place for safety, or so I was told when I visited that museum.

But the cast is equally interesting, showing two long lines of tracks—one made with big feet and the other with smaller feet—so close together that the two may have "had their arms around each other." Or that's what's said, although to me it seems more likely that the one with smaller feet was following slightly behind the other, a little to the left. From experience with pre-contact hunter-gatherers (which I'll talk about later), I noticed that they always walked in single file, leaving tracks like those in the museum, but they never for any reason put their arms around each other. Try it yourself. You can do it if you're on a sidewalk, but you're sure to stumble if you do it on uneven ground. Even so, although putting an arm around someone else while walking seems modern, and only done safely on a sidewalk, this doesn't mean it couldn't have happened in the past.

The famous Lucy was an australopithecine. She died about three thousand years after the two who made the tracks, and was a hardworking female, maybe four feet tall, weighing about sixty pounds. She spent her days finding food for herself and her children, if she had any, or maybe for those of her co-wives if she didn't. One theory suggests that australopithecines lived by the gorilla system, whereby several female gorillas and their children live with a big alpha male.

Lucy was killed by a large, lion-like cat, perhaps a giant known as *Dinofelis*,[*] who perhaps grabbed her from behind while she was squatting, trying to dig up an edible tuber. She would have known the danger of squatting but would have had no choice if she needed the tuber. Long, lion-type teeth pierced her pelvis, which was learned from tooth marks on her scattered bones. I hope she was dead before that happened.

It's important to remember that the hominids (the great apes including us) and the hominins (our ancestral branch of human types) spread widely, morphing in different ways to suit their environments. We tend to picture our ancestors as if they appeared one at a time, probably because we believe that we're distinctly different (in our own eyes) from the other great apes. But actually, several kinds of proto-humans would have existed together, so when one thinks of early human-types, it helps to consider other kinds of animals, each kind with different

---

[*] *DYE-no-FEE-lis*—"terrible cat."

species, many of whom live in the same places at the same time. For instance, cougars, lynxes, and bobcats live in the same areas if not always the same ecosystems. Foxes and coyotes live in the same forests. Jackals and African wild dogs live on the same savannahs. All kinds of animals do this, and our kind did this too.

Two million years ago, as Lucy's kind was disappearing, *Homo habilis* ("skillful man") appeared in the area of Olduvai Gorge in what is now Tanzania, one of the first of Genus *Homo*. A lesser known hominin, *Homo rudolfensis* ("man from the Lake Rudolf area") appeared a little later, also in what is now Tanzania. His teeth were different from those of *Homo habilis*, suggesting different diets, so they may have lived in different kinds of ecosystems, but were probably aware of each other.

The ancestor of *Homo habilis* also gave rise to *Homo erectus* ("erect man"). From a human point of view, these hominins had considerable abilities. *Homo habilis* made stone tools—thousands were found in Olduvai Gorge—possibly the first tools made by humans unless his recent ancestor had already made a few. So many stone tools were found that it almost looked as if those proto-people were running a business. Even so, *Homo erectus* is credited with making the first stone hand ax and also was the first to use fire. Already our kind was inventing—a process which we as their descendants continue. We're not the only animals who use tools—far from it—but we made our tools instead of finding useful objects (although some animals alter their found objects), and we're still the only animals who use fire.

Homo erectus could be considered our most conspicuous ancestor. His kind, like *Australopithecus* and *Homo habilis*, first appeared in East Africa and were so successful that, during the ten thousand centuries of their existence (nine times longer than we've existed), they moved out over Africa into Eurasia, modifying as they spread. Perhaps the best way to consider this segment of our evolutionary past would be to view some of the later modifications not as different species, which some taxonomic names would suggest, but as ever-evolving forms of *Homo erectus*. After all, these hominins led straight to us.

Homo erectus stood erect. If not, they would have called him something else. *Homo erectus* men were big—some of them maybe six feet tall. But the women were smaller—perhaps five feet tall. To have large males and mid-sized females is normal enough for the great apes, including our branch of them.

According to one theory, the presence of certain animal fossils associated with those of *Homo erectus* suggests that the men ran antelopes down. The bigger you are, the better you can do this, as your lungs are deep and your strides are long. But perhaps it was the determination of these runners, not their size, that helped them. Not that the following is diagnostic, but if some descendants of *Homo erectus* had kept up the practice, the descendants in question would be the San or Bushmen of the Kalahari—the first of us humans—whose men also ran antelopes down. San men were not much more than five feet tall and only slightly bigger than San women, so all you really need to do this is endurance.

Several kinds of human-types were present with *Homo erectus*, certainly *Homo habilis*, and possibly others as well, and at one point, maybe a hundred thousand years ago or maybe earlier, some *Homo habilis*-types managed to cross the ocean to the volcanic island of Flores off the coast of Indonesia. How they did this is unknown (enter the floating debris theory), and once there they did not become larger as did some other human-types, but may have become smaller, as often happens with an island species. Their descendants, *Homo floresiensis*,* were maybe three to four feet tall and weighed maybe fifty pounds—just a tad bigger than Millennium Man, and just a bit smaller than *Homo habilis*.

An extinct species of elephant, the stegodon, also lived on Flores Island and also was small, at least for an elephant. And although this was an interesting twist of evolution, it wasn't unique, because island species tend to be small. The reason isn't perfectly clear but could be due to limited food and few predators. The smaller your surroundings,

---

* *HO-mo floor-EEZ-ee-EN-sis*—"man from Flores."

the less food you find, and the smaller you are, the less food you need. And if no one else lives there who eats the smallest of you, pushing you onward to a larger size, you stay small. It happens in much of the Animal Kingdom and is known as "insular dwarfism."

These little people did well—living undisturbed for fifty thousand years. Then they disappeared, perhaps because the volcano erupted (one theory) or because our kind arrived on Flores Island and eliminated them (another theory). If this hadn't happened, Genus *Homo* might now have two species, and we might have known them. But if those little folks were here today, we might be as awful to them as we are to our other primate relatives. We eat all the others—the great apes are a favored form of bush-meat, offered in certain high-end African restaurants—so perhaps we would eat them too. And even if we didn't, we already call them "hobbits" and would see them as a major curiosity, putting them in cages and keeping them in zoos where we could taunt them and throw peanuts at them. We certainly would have taken their island from them. Surely something was on it that we'd want.

As for the *Homo erectus* people on the mainland, one wonders about their social arrangements. Did they live in large groups like chimpanzees, or in mid-sized groups like bonobos, or in smaller groups like gorillas? It's unlikely that they lived without groups like orangutans, because they seem to have been intensely social and are thought to have cared for their older members and those who were sick or weak. They are also thought to have made "expressive vocalizations," or in other words, they talked. Of course they did. Many animals talk; why wouldn't they?

We know they made stone tools and understood fire, the use of which may have come at least a million years ago when a *Homo erectus* realized that if lightning started a wildfire, he could ignite a dry branch to carry somewhere and ignite something else. Fire became especially useful to those who lived on the ground and is often believed to have discouraged predators. The theory was perhaps formed by people who live in buildings and don't realize that if predators then were like predators now, they also understood fires and would have been indifferent to small ones, especially small, contained ones such as campfires, because they knew they could avoid them. Even so, if you shake a burning branch at a predator, he may back off.

Our *Homo erectus* ancestors used fires long before they figured out how to start them, relying on lightning and wildfires to get them going. One wonders why it took them so long, when starting a fire is relatively simple once you know how. On the other hand, you don't need to start one if you keep one going.

In later years, the San also used a burning branch to start another fire, even though they had no problem starting fires. But doing so can sometimes take five or six minutes from the time you start until a fire's going. It's easier to keep some burning coals handy. Among the San, months could pass before anyone needed to start a new fire (and only then when people were traveling or moving to a new encampment), which a man always did by twirling a male stick in the hole of a stationary female stick, which he held down with his foot and in which was a little pile of tinder. The process went faster with two men twirling, but it always took plenty of time.

Perhaps control of fire was the most important discovery ever made by humankind. We think of such technical advances as the work of scientists in laboratories, but this one was made by a guy in a leather loincloth sitting on his heels under a tree. Maybe he dropped a stone on another stone, noticed a spark, and considered the potential. Maybe he was twirling a stick against another stick for no particular reason and saw a wisp of smoke. We assume such discoveries are made by men, but we have no support for that. The discovery could just as well have been made by a woman.

An African species to start with, most of the *Homo erectus* people stayed in Africa. But some of them—surely those in North Africa—eased into the Middle East, then on across Eurasia. Their fossils have been found all the way from Europe to India and China. The famous Peking Man was a *Homo erectus*. He lived in China perhaps 750 thousand years ago.

We imagine such travels as purposeful migrations but generally this cannot have been the case. If we felt secure and supplied, we would have had no reason to move. Our only incentive would either have been overpopulation—not a common problem in the natural world— or because a condition arose that we couldn't handle.

Perhaps our water disappeared. Perhaps something diminished our food supply, or our predators became more numerous, or new predators arrived, or groups of our own kind were threatening us. Our response to such events might very well have been to move, but no farther than we had to. Our survival depended on our knowledge of a certain kind of area. To move to a different kind of place—from a savannah to a forest, say—would have been a major challenge that we would only have accepted if we didn't have a choice. And even then, we would have eased into the new environment rather than marching in boldly. This must be why our move to other continents took thousands of years, and was accomplished by many generations. Even if migrating groups moved only four miles a year, it would have taken them no more than six thousand years to go all the way around the world. It took most of them longer than that just to reach the Middle East.

The slow spread of *Homo erectus* took such a long time that in various locations they'd change somewhat to become *Homo this* or *Homo that*. One of the African forms, *Homo idaltu* ("first born Man"),* seems to have been an early form of *Homo sapiens*. Another form, *Homo heidelbergensis* ("man from Heidelberg"), appeared in Eurasia and may have become the Neandertals ("those from the Neander Valley"). But *Homo idaltu* and *Homo heidelbergensis* were almost *Homo erectus*, because the DNA that passed on to us and the Neandertals differed by only 0.12 percent. This means that the difference between the African and Eurasian species was slight—something like the difference between lions and tigers. These cats can breed and produce fertile young, and so could we and the Neandertals.

---

* From Saho-Afar, an Ethiopian language.

As for us, the *Homo sapiens*, we began in Africa, and most of us stayed there. But we had been moving around as almost any species is inclined to do, and although the following dates are uncertain and different versions are offered, at some point, maybe about forty thousand years ago, a few *Homo sapiens*—according to one theory, perhaps no more than a hundred and fifty people and probably not all at once—went to the Near East just as our *Homo erectus* ancestors had done before us. We spread slowly northward from there, and in Europe we met the Neandertals.

CHAPTER 21

Neandertals

Neandertals and ourselves could almost be called evolutionary cousins, but after our arrival in what is now Europe, our contact with them is uncertain. We are sometimes told we didn't get along with them and are to blame for their going extinct. But that might be because we now are deeply involved with fighting and assume that most lifeforms fought and always did. Witness the assumed war zone of the dinosaurs—the "Terrible Lizards"—occupied by Tyrant Lizards, Monstrous Murderers, and Kings of Gore, and the Pouched Lion Executioners and the Dinofelis lion-types, the "Terrible Cats." We use the word "fight" for the efforts we make; we fight cancer, crime, and climate change, and we're prone to physical fighting. At the time of this writing, wars are in progress in Afghanistan, Bahrain, Bali, Iraq, Israel and Gaza, Libya, Nigeria, Pakistan, Somalia, Sudan, Syria, the Ukraine, Yemen, and Zaire, to say nothing of drug wars in Colombia, Honduras, and Mexico.

Thus we tend to assume hostility when two groups collide. The Vikings, for instance, were assumed to have swept down the coasts of England, raping and pillaging, when all along, many were marrying and settling, hence the British blondes. As the Vikings were to the British, so may we have been to the Neandertals, but not as scary. We're speaking of two far-apart time periods, of course, so this is just for comparison, but when a Viking ship appeared off British shores, it was powered by forty or fifty large-size men who hadn't seen a woman for several months, whereas when we appeared on Neandertal lands, if we were anything like the present San, we would have been fit and strong but short and slight, and would have come on foot in family groups, our men with women and children.

A Neandertal man was massive if not very tall, perhaps measuring about five and a half feet, but might have weighed a hundred and eighty pounds because he was so muscular. His kind could deal with cave lions and two kinds of giant bears, so it's hard to imagine a few *Homo sapiens* presenting much of a problem. There were more Neandertals than there were of us, certainly at first, so if our men had taken their women or otherwise caused them serious problems, we might have been the ones who disappeared.

Instead, we coexisted with Neandertals for twenty thousand years. Why they disappeared hasn't been determined as far as I know, unless we blended with them. Considering the time we spent in relative proximity, it's hard to imagine how we coexisted if we often engaged in serious friction.

Unlike the African savannah where our ancestral humans lived in shelters made of grass and branches, much of Eurasia was forested. The terrain of much of Europe, especially near rivers, lends itself to caves, and many caves have signs of early habitation. That's because we lived in them, having winter and predators to deal with. A cave not only prevented predators from sneaking up on us but also kept us out of the wind and relatively warm, retaining heat from a fire that we made ourselves rather than waiting for wildfires.

We once called our ancestors cavemen and also Cro-Magnons, named for the cave where the first fossils were found and also for a hermit who happened to be living in that cave at the time. Now we call them "European Early Modern Humans," which seems like political correctness although it's hard to see the need. These people left their magnificent paintings in many of the caves, some of which were occupied by our species from the time we first arrived until well into the twentieth century.

As recently as 1949, people were still living in the caves above the Dordogne River in France. When my mother and I visited the Dordogne Valley that summer, we saw laundry drying on a clothesline outside one of the caves. But no one lives in these caves today, and if the residents were evicted it would probably have been for the reason described by a fascinating article in *The New Yorker*, "A Cave with a View" by D. T. Max. He described the caves of Matera in southern Italy that had been in continuous use from the Paleolithic to the 1950s, but after that, their use was considered disgracefully primitive, and the residents were forced to leave. "The transfer of Materans is seen as one of many patronizing attempts by élites to save indigenous people from themselves," writes Max.* These caves are still in use, but now as tourist attractions.

Some of us have Neandertal genes. Neandertal genes do not appear in those of us whose ancestors stayed in Africa, which means that those of us in Europe intermarried or at least interbred. But a mystery remains. Although some of us have Neandertal DNA, none of us seem to have their mitochondrial DNA, known in science-speak as mtDNA, which is transmitted only by the mother.

Not everyone has been tested, of course, but it's a fascinating thought. What can explain it? Could it be that Neandertal men impregnated our women while successfully defending their women from our men? Could this have held for every romantic encounter during the twenty thousand years we coexisted? Or were our interspecies matings such that when the mother was Neandertal and the father was *Homo sapiens*, her children were infertile? This sometimes happens

---

* D. T. Max, "A Cave with a View," *The New Yorker*, April 27, 2015, 36–37.

when different species mate. Witness mules—their mothers are horses and their fathers are donkeys, and the mules that result are almost always infertile females. The third possibility is that many of us did get Neandertal mtDNA which has vanished due to random changes. This happens often enough to make this possibility the most likely.

There's yet another possibility, however faint, based on the suggestion that Neandertal men had heavy body hair and massive beards. Our species began in a hot climate, and so when we arrived in Neandertal lands we had very little body hair and no beards to speak of. What if our women favored hairy bearded men? A beard would show that the guy was a grownup and likely to be a better hunter than a boy, which might explain it, but Darwin believed that men acquired beards through sexual selection. Women eagerly select men with beards, he said. He had a beard, so maybe he spoke from experience.

As for me, I would have liked Neandertal men. A heavy-set, bearded guy who could hunt down just about anything, carry it back to the cave on his shoulders, and build a fire to cook the meat? What more could you ask for if you were a woman hunter-gatherer? Or any kind of woman, for that matter?

But when I think of the Upper Paleolithic, I seldom think of Neandertals. Instead, I think of a cave painting my mother and I saw of two lionesses, one on either side of an opening to a deeper part of the cave. They're sitting down, looking at the viewer (or at least your eyes meet theirs when you come around the corner), and are so well done, so accurate—not only their bodies, but also their demeanor—that they took my breath away. Why are they on both sides of the opening? Are they guarding it? That's what it looked like, but guarding it from what? The paintings were the work of a Cro-Magnon individual, a *Homo sapiens* like me. I wish I could have known him. I wish I had seen the lionesses.

So I imagine myself by that cave long ago, when the artist lived inside it. I think of the cave as I remember it, on a long, sloping bank of the river that's running below it, and suddenly I'm there. The cave yawns above me, and a bearded man wearing a leather loincloth is standing in front of it. By chance I'd materialized in a cluster of bushes, and when I step out he is shocked. He utters two bark-like words with a questioning tone. I can't understand him so I smile and look down at my hands. "I'm harmless," I'm saying.

This is not enough. He turns his head and calls into the cave. Another man hurries out. He's frowning. The two of them stare as if deeply concerned. I lower my chin and look up at them, my arms down, my palms out, trying to say "I'm nothing." This only starts both of them barking fast in loud voices as if to me, to each other, and to whoever is listening inside the cave. I see that I'm a disturbance, so I take a step backward and turn to move farther away.

It's then that I see the lionesses. They must have been approaching the cave when I came. When they see me they crouch, tails twitching, ears low, eyes wide. *This can't be good*, I tell myself. *I'm going home*, and I'm back in my house, breathing heavily. But in my nostrils is the soft, musky scent of the lionesses, and in my ears is the sound of men shouting in commanding tones that boom back as echoes from the cave. Did the lionesses want to eat the people? Did they want to take over the cave? What happened after that I'll never know. The painting of two lionesses remains unexplained. I should have climbed a tree and waited.

CHAPTER 22

Why Do We Look the Way We Look?

Why do we have no fur when fur protects us? Mosquitoes can't bite through heavy fur, or not easily, and African nights can be freezing. Bare skins are for those such as rhinos and elephants who need to lose heat, not keep it. And why do we walk on our hind legs when walking on all fours is easier, faster, and safer?

One theory holds that we ran on our hind legs to escape our predators, but that's unlikely. Any mid- to large-sized animal runs faster than we do, and when we're running from something, we present our backs. Roll a tennis ball across the floor and see what your cat does. To run from predators invites them, so that can't be why we move the way we do.

As for standing up straight, if we do we see farther than we can when on all fours. We also look bigger, and looking bigger is a widespread defense. It's why bears stand up on their hind legs, and why our skin prickles when we encounter something scary. When our bodies were covered with an important amount of hair rather than the pitiful little strands we have today, we could make ourselves much bigger by bristling. If a mid-sized predator approached and we stood up and bristled, he might think, *Uh-oh, not this one,* and leave. Yet despite all the theories, the long and short of this question is, no one really knows why we walk on our hind legs.

Why we lost our body hair has also inspired theories. We were an African species whose ancestors lived in shady forests but later came to live on savannahs where the days can warm to 120°F or higher. Perhaps we needed to cool ourselves by shedding our body hair, but kept our head hair as protection from the sun. That's one theory, but it's worth remembering that baboons still have hair on their bodies as well as on their heads. Why didn't they lose their body hair and walk on their hind legs? If they didn't need to, why did we?

Another theory suggests that we kept the hair on our heads to shade us from the sun but shed the hair on our bodies to get rid of lice. Then we were over-exposed to sunlight, so we made clothes.

If we had also shed the hair on our heads and made hats instead, things would have gone better. But we didn't, and some of the lice who stayed on our heads evolved into body lice who live in our clothes. This could date the invention of clothes, which seem to have appeared along with lice about a hundred thousand years ago. Our clothes created whole new ecosystems with us inside them, and it only got worse when some of the lice in our clothes evolved as crotch crabs.

Thus the invention of clothes didn't bring the benefits we had imagined, because after that we had head lice, body lice, crotch crabs, *and* clothes which we had to clean and mend. If we'd thought this through before shedding our fur, we'd have only one of those problems.

Our louse misfortune casts doubt on the creation theory. If everything was created all at once, Adam and Eve were created with three kinds of lice plus tapeworms. "Why all these pests?" they would have cried, furiously scratching. No wonder they left the Garden of Eden—no doubt to plunge into the nearest pond, where guinea worms, Giardia parasites, and snails that carried snail fever were waiting! But who can say? Perhaps the Creator favored the parasites and made humans to be their food.

It seems risky to include clothing as an evolutionary indicator. Until very recently—only about sixty years ago—our direct ancestors, the San, had traditionally worn minimal clothing. Men wore leather loincloths, and women wore front and back leather aprons and leather capes as pouches for their babies. The men were excellent hunters, so their people had a steady supply of antelope skins from which they could have made all kinds of clothes. But they didn't. Skin pigment protected them from the sun; their leather capes and campfires protected them from the cold; and they didn't have lice, although their minimal clothing could have harbored them.

The above is just one small fact to be learned from the San, who are known as the first people. According to DNA studies, the rest of us descend from them, and their languages seem to be the root of all languages. Living in what I've named "the Old Way," they could show us more about life in the natural world than most scientists, not only because they knew every inch of their part of it, but also because knowledge was the key to their survival. The San showed what knowledge it takes to live in the natural world, the kind of knowledge all species who do so must have, in one way or another.

CHAPTER 23

The San, Formerly Known as Bushmen

Beginning in 1949, my father, Laurence Marshall, organized a series of expeditions in which he, with my mother, Lorna Marshall, my brother, John Marshall, and several others including myself, then Liz Marshall, visited an "unexplored" area, about 120 thousand square miles in the northern part of South West Africa (now Namibia), the western part of Bechuanaland (now Botswana), and the western part of South Africa, including most of the Kalahari Desert. White people, who then controlled South West Africa, called it "the end of the earth."

There we met the San, then known as Bushmen, in an area of about six thousand square miles they called Nyae Nyae. The people were pre-contact, or in other words, they were untouched by what we love to call "civilization." Nor did they have the diseases that now infect the so-called "developed world." Anopheles mosquitoes were present, as were the snails that carry snail fever, but the San we knew were in perfect health and didn't get malaria or snail fever because only the vectors were present, not the parasites themselves. "Pre-contact" is often seen as a disadvantage, but this was far from true. They lived as our species had lived for at least a hundred thousand years, and their culture may have been the most successful culture the world has ever known.

I use the term "San" rather than "Bushman," because "San" is now preferred and is also official. I put quote marks around "unexplored" because, as archaeologists later discovered, one of the San encampments had been occupied continuously for more than thirty thousand years and another for more than eighty thousand years, both showing little change in the material culture. The San had done the exploring.

To know these people has value. Not only did they show us how our species survived, they also showed us the kind of information that any life-form needs to do this, each in its own way. To some extent, this shows our commonality with other life-forms, so I'll tell about a strange event—apparently minor but very revealing—to prove it.

I once was with two San men who, evidently for no reason other than curiosity, were following the tracks of a hyena. And as they did, they looked ahead of themselves, now and then glancing down at the tracks.

The hyena had walked through some heavy brush and emerged on an enormous space of bare rock, perhaps a quarter of a mile long and an eighth of a mile across to another area of heavy bush. The hyena had been walking in a fairly straight line, let's say toward twelve o'clock, and then had stepped out on the rock. No way would his footprints show on the rock—I assumed there'd be no more tracking.

But the two men kept walking. They went out on the rock and without changing pace began curving to the left, let's say toward eleven forty-five. I followed, of course, and soon enough we were in the bush on the far side where once again the hyena's tracks were visible. It seemed like one out of dozens of places where a hyena might want to go.

How did those men know where to follow? They knew the hyena's mind. When they scanned the bush on the far side of the rock, which looked all the same to me, they instantly saw the kind of place a hyena would choose to enter. No problem.

They also knew the minds of every other animal in that 120 thousand square miles of bushland savannah, a skill they obtained because they'd noticed that we and other animals are much the same.

It hurts a bit to speak of humans as animals, but of course we are and always were. Everyone from Darwin to a housefly is or was an animal. That's why, if we know enough, any other mammal is more or less predictable. Our ancestors had this kind of knowledge when they came out of the trees. Anyone *could* do what those men did while following that hyena, but those men had the necessary knowledge, and the rest of us don't.

At the time of our first visit, most white South Africans thought of the San as a kind of wild animal. Some white men with rifles hunted them for sport, hiding near a waterhole and shooting whoever came to drink. The San in the interior were known as "wild Bushmen" and weren't considered landowners because, as was said, they "didn't use" the land. Every time a new farm started near the edge of the "unexplored" interior, the farmer took the waterhole if there was one, offering the San who lived there the choice of being homeless or working for him. Those who stayed were given cornmeal to eat, maybe a shed to sleep in, and a few ragged clothes, but no money.

The San were grossly exploited, but they were also feared. They lived like animals, after all, so they must have been aggressive. While we were in Windhoek—the capital of what is now Namibia—where we prepared for our journey, we were told that Bushmen would shoot us with poisoned arrows. But during all the years we spent among these people, we were treated with respect and kindness, possibly because that's how we treated them. The San were misunderstood.

In fact, the San knew the dangers of aggression better than most and dealt with it in many ways because it compromised their chances of survival. From their personal interactions and their conduct in daily affairs to the structure of their social systems—kinship, marriage, various kinds of partnerships, and the like—essentially everything they did reflected their value of cohesion. So did their name for themselves. The people we knew best called themselves Ju/'hoansi. With help from our interpreter, a wonderful multi-lingual Tswana guy named Kernel Ledimo, I translated *Ju/'hoansi* as "harmless people." *Ju* (pronounced *zhoo*—the *zh* as in *azure*) means "person"; /'*hoan* (pronounced *hwa*) means "clean, safe, not harmful"; and *si* (pronounced *see*)[*] makes a word plural. The / is a click you make with your tongue popped down from the top of your mouth, and the ' is a tiny little pause.

Years ago, I wrote a book about the San called *The Harmless People*. Some academics were critical because they could not believe there was such a thing as a non-aggressive human population. One famous biologist—having heard that our family's research showed the Ju/'hoansi as valuing cooperation and discouraging aggression—claimed they were controlled by the police. This from a scientist? During those thousands of years in those thousands of square miles known as "the end of the earth," pre-contact hunter-gatherers didn't have police. The australopithecines and the *Homo erectus* proto-people probably didn't have police either, so that isn't the explanation.

An academic who wrote about our experience told me she'd read *The Harmless People*, but I think she missed the part about the Ju/'hoansi having chosen that name. She thought the "Harmless" in the title was my invention and showed we had a romantic view of hunter-gatherers. This caused us, she thought, to underplay the fact that the Ju/'hoansi had weapons—in one of our publications we included only one photo of a man with a spear.

But a spear is to a poison arrow what a butter knife is to a gun. No matter where the arrow hits—even in your foot, even if you pull it right out—you die if the poison gets in you. A bullet isn't that good, let alone a spear. Yet all our publications include plenty of photos of men with bows and arrows. This seemed to stir no irritation, because everyone knows that arrows are for hunting. All of this goes to show

---

[*] These pronunciations are approximate, needless to say. The *hwa* has a lowered, ringing tone at the end, hence the *n* in the written version.

that our critics were so convinced of hunter-gatherer aggression that they scrambled for evidence wherever they could, while dismissing our claims to the contrary.

But who can blame them? As prisoners of our own culture, we see the natural world as red in tooth and claw. Wild animals are seen as aggressive when hunting, although if they were people they'd be shopping, and if a lion is roaring or a coyote is howling, the noise is seen as a threat. If that's how wild animals deport themselves, how could wild people be different? Maybe if the San didn't hunt but only gathered, the people who thought they were aggressive would see them as peaceful and sweet.

I do think that hunting is vaguely to blame for these misunderstandings, if perhaps not as much as living in the wild. But however mistaken the concept of aggression may be, it isn't new, and it applies to almost anything that hunts. Our australopithecine ancestors were assumed to have been aggressive fighters who killed and ate not only each other but also the large lions that lived among them, thus contributing to their extinction. Although the australopithecines were little more than four feet tall, the concept was accepted by certain paleoanthropologists, evidently including none other than Louis Leakey, who said that when it came to animal predation, australopithecines were "not cat food."* He was mistaken, as we now know from hominin fossils found with lion tooth marks on them. Abundant evidence shows not only that humans were the prey of large cats and sometimes hyenas, but that it goes on to this day. We've been the meals of such cats for millions of years continuously. For as long as we've kept any kind of records, lions have been man-eating, as have tigers, leopards, jaguars, and cougars. Australopithecines were not cat food? Could a paleoanthropologist think that man-eating was new?

Then too, is man-eating aggression? Yes, if you are the human; no, if you are the cat. And as for humans defending themselves, is that

---

* Marcus Baynes-Rock, "Mark of the Beast: Reflections of Predators Past in Modern Mythology," Academia.edu, 2.

aggression? No, if you are the human; yes, if you are the cat. Noting the mixtures that constitute aggression, it's all in your point of view.

Today the San live as rural Namibians, and their former way of life is gone. But plenty of people want to describe how they lived it, and in doing so they promote misunderstandings. The San were said to be superstitious. They didn't go to school, thus they never learned anything, so they made up stories to explain the world. Some anthropologists seemed more interested in their legends than in their actual knowledge. But we found their knowledge breathtaking. They had named and knew the properties of virtually every visible life-form in their ecosystem. As for natural events, I once asked one of the men where stars went in the daytime, perhaps expecting a story, and I will always remember his answer. "They stay in the sky," he said. "We don't see them because the sun is too bright."

During a total eclipse of the moon I asked another man what was happening. "Don't worry," he said. "The moon comes right back." That's a long way from superstition. But it doesn't mean the San had no stories. Someone else said a lion had covered the moon with his paw so he could hunt in darkness.

We're willing to see this last remark as evidence of superstition even though we have our own stories. Our Christmas presents are under the tree because Santa came down the chimney. What we don't appreciate is the underlying truth of the lion and his paw, as the San had always known what most non-San people still don't know: that a lion gets hungry on moonlit nights because his potential victims can see him. Of course he wants to cover the moon, especially if it's full, as it always is when in an eclipse. He hunts better in darkness.

One author wrote that the San found their food "intuitively." In a different version of the same idea, another author wrote that the San went wandering around the savannah in search of food and water. Hunting might be vaguely described as "in search of," but if people had been living in the same place for eighty thousand years, wouldn't you think they would have found the water?

For us, such lack of perception seems normal. I once did a little study of our awareness by questioning several college students about the sun.

Q: "Where does the sun rise?"

A: "Huh? Um..."

One young man pointed to where the sun rose that morning. He didn't mention east, but so what? He had actually looked at the sun and remembered where it came up. That's all you need to do to echo the Ju/'hoansi, but most of us don't bother.

As for the Ju/'hoansi, they had acquired an overwhelming amount of astronomical, botanical, zoological, and climate knowledge, which few have equaled to this day. This came from flawless observation, as it did and still does to all who live in the natural world. The Ju/'hoansi knew and had named just about every kind of plant and knew their properties. They also knew every kind of animal, including insects, and their behaviors. So when a Harvard professor, after visiting the Ju/'hoansi briefly, said they "know almost as much as we do," he fell short. They knew more than we do, but if you know less than the person you're questioning, you won't know what to ask.

The professor didn't know to ask about a weaver bird's nest, for instance. It's a long, bag-like nest with lots of fibers, and according to the San, it has a pocket for the nestlings and a pocket for a snake. Educated people know that a bird would not accommodate a snake, so the professor would have seen this as a superstition. But as the San had long ago discovered, when a snake climbs the nest in search of nestlings, her weight pulls the nest down, opening the snake pocket and closing the nestling pocket. The snake goes into the open pocket, finds nothing, and leaves. The professor couldn't have asked about something he didn't know existed.

One day a young boy named /Gao showed me hyena tracks he'd seen near the encampment. He said the hyena went by just before sunrise. The night before, we'd heard two hyenas whooping and hyena-laughing near the encampment, but they stopped around midnight. Did they stay after that?

I asked /Gao how he knew when the hyena made the tracks, and he showed me. In one place, the tracks were pierced by tiny pinpoints which I could barely see even after /Gao crouched down and put his finger by them. The hyena's tracks were fresh, he said; the tiny tracks were made by a dung beetle, and the beetle would not have been moving until the sun was up to warm the air. If the beetle had come long after sunrise, the dust would have been too dry to show his tracks. Such tiny, delicate tracks would appear only in certain conditions. Therefore, the hyena walked by before sunrise, and the beetle crossed his tracks soon after that.

How's that for knowledge and observation? This from a boy eight or nine years old who happened to walk by a line of tracks, glancing at them in passing.

Not only did I learn about the beetle from /Gao, I also learned about hyena reproduction. After seeing the tracks, I sat in the encampment, doing nothing in particular, just watching /Gao and another boy playing with a kite they'd made from a feather. For some reason, the other boy made a hyena whoop. I asked the boys if they'd seen the hyenas the night before, and they said they did. I wondered aloud what the hyenas were doing, and /Gao told me they were mating. The quarter-moon had given light, and evidently the boys had seen them. I wished I'd seen them too, and soon I might as well have. The two boys did a reenactment.

At first, the other boy played the part of the female hyena and /Gao played the part of the male, but soon they saw they were doing it wrong, so they swapped. And rightly so—/Gao was taller than the other boy, and female hyenas are bigger than males. The "male" would mount the "female," and the "female" would struggle to heave him off. They'd turn and look at each other, then tussle for a while and try again, whooping, growling, and hyena-laughing, sounding just like the hyenas did the night before and taking the same amount of time to do it—maybe ten minutes. The "male" didn't actually penetrate the "female," he just pretended to, humping as he did, but aside from pretend penetration, the reenactment was faithful. No one could make this up.

It had seemed to me that the real hyenas fell silent around midnight, presumably because their mating was successful, so why one of them waited to leave until just before sunrise was a mystery. This seemed like a complicated question and I didn't speak Ju/'hoan well enough to ask. Perhaps it was a different hyena to whom the mating of the others was of interest—I'll never know. The take-home is that two young boys had flawless information about hyenas, and one of them noted tiny pinpoints from a distance, knew they were tracks, knew they were made by a beetle, knew what kind of beetle, and knew the time this happened. All that was as casually normal to him as looking out the window is to us.

Surely the most impressive evidence of San observation was their arrow poison, perhaps the deadliest in the world. One drop can kill you, and there is no antidote. It's found in the larvae of *Diamphidia* beetles and their *Lebistina* parasites—these two among thousands of kinds of African beetles and parasites. The grub hatches from an egg laid by the mother beetle on the leaves of a *Commiphora* tree, which grows in sandy soil. The grub crawls downward under the bark, exits the tree through one of its roots, and makes a cocoon in which to pupate by sticking grains of sand all over his body. He looks like a tiny ball of sand and thus is inconspicuous because around him are similar tiny balls or lumps made only of sand.

The adult beetles are not impressive—they're small and dark-colored, and if they're present at all, they're just standing around on the *Commiphora* leaves. They never interfere with people, so you seldom notice them. Even if you happened to dig in the sand, the pupae are down in it two feet deep, but nothing much grows under those trees, certainly no food items, so you'd have no reason to dig there. Then too, if someone found a cocoon, removed the sand grains, opened it, and removed the grub, nothing would happen because the poison is inside the grub. Even if the person ate the grub, it wouldn't necessarily harm him unless he had an ulcer, because the poison must enter the bloodstream to do harm. If the arrow has killed an antelope, you can safely eat the dark-looking meat where the poison arrow entered and some of the poison still clings. Even so, long ago, someone discovered the grubs and the poison.

The poison itself is also far from obvious. It seems to be under the upper leg of one kind of grub (maybe the parasite's) so you pull off the leg and squeeze the poison from the opening. Poison must be all through the body of the other kind of grub (maybe the beetle's) because with these you mush up the body, pull off the head, and squeeze out the mush like toothpaste from a tube. And you do this very carefully, because the smallest drop of poison can kill you if it gets in your blood.

How people learned of the poison is as mysterious as finding the grubs. As I've said, to be fatal, the poison must enter the bloodstream, where it destroys the hemoglobin. Hemoglobin carries oxygen around your body while removing the carbon dioxide, thus as you start to lose it you feel dizzy, weak, and short of breath. Death doesn't come for several days, so if someone who was poisoned didn't die fairly quickly and at first didn't even feel sick, how would the other people know what killed him?

Yet long ago this discovery was made, as important to the hunter-gatherers as control of fire has been to the rest of us, and could only have been made by people who examined the natural world more carefully than few of us have ever done since, by people who were intelligent and thoughtful and needed to learn about what they were seeing, and thus knew almost everything there was to know about their environment that could be seen without a microscope.

We now reserve such knowledge for scientists. But for our San ancestors, learning began in childhood, and everyone gained the same knowledge. Those who say that the San were uneducated are applying to our immediate ancestors the same misconception they apply to other species. Ju/'hoan knowledge wasn't gained in schools and thus was not considered "education."

The San lived the same kind of life as does every other life-form in the natural world, quite different from ours and that of our domestic animals. If we bear in mind that different species have different needs and different ways of solving their problems, the San showed what it takes to do this. We've been thinking of other species as hard-wired, born with knowledge, governed by instinct, forgetting that our ancestors lived the same way. Our ancestors aren't seen as hard-wired, and unless we can prove hard-wiring in all the other species, we must accept that the other species also observe and learn. Let's not forget the Mimosa and the drops of water, or the "white rat" paramecium and the dangerous light, and learn from our own kind, the pre-contact San, the importance of understanding the environment, and configuring a lifestyle that contributes to survival.

CHAPTER 24

Gaia's Rule One

Find a Source of Energy

The lifestyle of the San was guided by Gaia's first rule: *You must find a source of energy to keep yourself going.* An enormous Ju/'hoan lexicon of facts on food-sourcing was passed down through the generations, so the knowledge of the San was awesome. I used to go gathering with some of the women, sometimes setting off long after sunrise to be sure the predators were sleeping, but sometimes setting off early, despite the predators, because the destination was far. We would walk for several miles to a place the women had chosen in advance, either because they knew that fruits there were ripening or tubers were ready to dig up, or that bushes with berries were in danger of being denuded by birds. Sometimes it could be all of these reasons.

We'd come to a berry bush, chase off the birds, eat berries for a while, and then move on until we reached our destination—maybe a small grove of trees with bushes and vines around them. The women didn't need to search for what they planned to gather. From a distance, if they were looking for a certain kind of tuber, they'd spot the tips of vines that grew from those tubers and would spread out to start digging.

They used special sticks to do this. These were made from strong, straight branches broken from trees and stripped of twigs and leaves. One end of the stick was pointed and perhaps fire-hardened. A woman stabbed this repeatedly into the ground until enough dirt was loose and could be scooped out with her hands.

Her toddler might be watching. She might show him the tip of the vine and name it. If she'd shown him this before, she might ask him to name it. He might be only three or four years old, but often enough he remembered and could name it. He'd then watch her dig until she'd exposed the top of the tuber. He'd watch her pull it out and brush it off. He'd watch her taste it to see if it was okay, and if it was, he'd watch her put it in the pouch of her cape.

Older children—those around five or six—already knew the names of the vines and would be playing actively but quietly nearby. When I sat by my friend who was digging, I seldom heard anything but the wind in the trees and her digging stick hitting the dirt. Except on the nights when they danced, the Ju/'hoansi were very quiet.

This would go on for several hours, until the women had gathered all they needed and would think of starting home. They'd call to each other and their children—their clear voices were surprising in all that silence—and then would start home in single file until they'd pass a fallen tree they would have known about, and would break off branches for their fires.

By then, each woman would be carrying almost her own weight—a child, a heavy load of tubers (the big ones could weigh maybe six pounds each), and the firewood. I'd help by carrying some of the wood or perhaps a tired child. We'd all be tired, but we'd step right along because we were still far away and the sun would be setting. Sometimes we'd reach the encampment just before dark, sometimes soon after dark, and sometimes long after dark. Never in all such trips I accompanied did we see a predator. But as night was coming we might hear jackals calling one another, or guinea fowls calling as they flew to their roosts in the trees. I seldom heard a large predator's voice, no doubt because the predators were hunting. They were out there, though, so we didn't talk.

Among the San, women provided the main diet and thus fed everyone most of the time, but meat was by far the favorite food and was always provided by men. They hunted, which as far as the Ju/'hoansi were concerned, was the most important thing that anyone ever did. They mostly hunted antelopes, but they also hunted giraffes and even Cape buffaloes when some of these came to the area during the rains. The irritation of the arrow made a dangerous buffalo more dangerous, but still the Ju/'hoansi hunted them.

Here, it's interesting to note that safari sport-hunters in other areas consider Cape buffaloes to be the world's most dangerous game. Sport-hunters use high-powered weapons—a .357 caliber rifle will do but a .458 caliber is better—loaded with soft-point bullets for the first shots to wound the buffalo and solid-point bullets to finish him off. The internet abounds with advice on killing Cape buffalo. Types of guns are described by men who pride themselves on having used them, often combined with stories about extreme danger and close calls that show the unbelievable courage of the hunter.

The Ju/'hoansi used little arrows that weighed about an ounce, shot from a bow with a twenty-pound pull, and didn't consider themselves to be courageous. When telling of a hunt, any kind of hunt, Ju/'hoan men told of the excitement of spotting the victim, how they stalked it and shot it, where the arrow hit, and what they learned from the victim's tracks as it ran away, but they never portrayed themselves as courageous. This was even true of a man who was gored by a Cape buffalo he'd shot with an arrow. He recovered, but didn't see himself as brave. His story made no mention of risk or danger.

Hunting was the main source of meat for the Ju/'hoansi as well as the source of skins from which they made their clothing. But it wasn't the only source of protein. The women who gathered plant foods also gathered animals such as tortoises and non-poisonous snakes, all known as "slow game." When a fifteen-foot python was seen drinking from the waterhole by the encampment where we stayed, an elderly woman killed it, threw it over her shoulder, and carried it to the encampment where she divided the meat among the children. But meat obtained by hunters was by far the most valued food, and hunting large antelopes was by far the most important activity that anyone could undertake.

Catching a python wasn't hunting. Women could have nothing to do with hunting due to their female power, which would weaken or dilute male power—a significant concept of the Ju/'hoansi. And why was this? Men and women had opposite powers. A man could hunt, providing the most valued food, and a woman could make another person. The powers were so different that they weren't supposed to mix. Women isolated themselves when menstruating and gave birth away from the encampment. After a baby was born, the mother would carefully bury all traces of the birth and mark where it took place so no man would go near it and encounter female power.

Because of this taboo, I never went hunting with the men. All I knew about hunting was what I was told, sometimes by the men themselves but more often by my brother, who regularly accompanied hunts. The men knew the habits and preferences of the various kinds of antelopes and would go to places where they'd be likely to find them. They'd watch the bushes carefully for motion. An antelope might be there, standing perfectly still because she knew the men were near. But maybe she'd flick an ear, and they'd see the motion.

They would stalk her very quietly until near enough to shoot a poison arrow—they almost never missed—and she'd go bounding away. The men would visit the place where she'd been. They'd examine the area and memorize her tracks. They'd also retrieve the shaft of the arrow, designed to drop from the arrowhead after it hit so the antelope couldn't grab it with her teeth and pull the arrow out. When the men had learned enough about the antelope, they'd track her.

Because the poison took so long to work, the tracking took several days during which the men might not eat. They didn't bring food or water with them because they already carried a bow, a spear, a knife, and a quiver full of arrows, and they'd only be encumbered if they had to carry more. Sometimes they found food along the way. In the dry season, finding water was a problem, but certain vines grew little watery melons, and certain hollow trees held water left over from the rains. The men could go for days without food, and chose water over food if they had to.

Their tracking skills were legendary. They'd know their victim's tracks if she joined other antelopes. If one of them left the herd, they'd know by her tracks if she was their victim. After several days, they'd come upon her lying down, sick from the poison.

Sometimes lions or hyenas would have found her first. On one occasion my brother was with four hunters when they found their antelope, sick from the poison but still alive, with a lion and maybe nine or ten lionesses around her. The men were seriously outnumbered, but they spoke to the lions respectfully, telling them the meat was theirs and the lions should leave. After pacing back and forth and grumbling a little, the lions obligingly walked away. My brother has this on film. One of the men then speared the antelope to kill her, then they skinned her, cut her up, and carried home the skin and meat.

The victim didn't belong to the hunter. It usually belonged to someone else, perhaps to the man who owned the arrow that killed it, which he would have given to the hunter. This man might be old or in some way disabled. If he couldn't hunt, he would never own meat if not for this custom. He, not the hunter, would divide the meat among various people who in turn would divide their shares among their relatives. This meant that everyone in the encampment got a share and would share again after the meat was cooked. The people loved meat. A party atmosphere prevailed on these occasions.

Girls accompanied their mothers while gathering, but boys didn't accompany their fathers while hunting. For one thing, hunting was too arduous. And a hunt could be ruined by an eager youngster. Boys had to learn the skills of hunting before they could accompany the men. But they were enthralled with hunting so they made toy bows and arrows and hunted dung beetles, present in all encampments (and the reason the encampments were so clean—if dung was nearby, a beetle would roll it away to a place where she could lay her eggs in it). The toy arrows were just sharp twigs, not poisoned, but enough of them would kill the beetle, and the excited boys would jump up and down in triumph.

From childhood to old age, the Ju/'hoan men were obsessed with hunting. They talked about it all the time. Yet hunting was so rigorous, so difficult and demanding—many miles of walking, days without food or even without water, tremendous skill and energy required, and a heavy load to carry home—that the men only went hunting every few weeks or even less often.

The meat of an antelope would feed an encampment of maybe thirty people for several days, but however much it was valued, it wasn't the main diet. As I've said, the main diet was vegetable food, and all but the very young and very old would gather this. Women did most of it, but men would gather if they felt like it. Whoever found a food item could eat it without having to share. Anyone could also find slow game—a snake or a tortoise, maybe—so even this didn't need to be shared. Like the python mentioned above, it was often given to the children.

The rules for sharing the meat of game antelopes were different because only those with skills and stamina could obtain it. Here, sharing was compulsory and formal. Without that, only the best hunters or the strongest men might always be the owners of the meat, and this could cause friction. Friction would lead to failures of cooperation, and failures of cooperation would cause the whole group to suffer in the end. With everyone observing Gaia's first rule fairly, in which finding yourself a source of energy could mean sustaining the entire group, nourishment for all was an important means of survival for the Ju/'hoansi.

CHAPTER 25

Gaia's Rule Two

Protect Yourself

Because of the value they placed on unity, the San adhered to Gaia's second rule concerning self-protection. Their division of meat, their kinship system, their marriage system, their partnerships with people far away, their system of sharing, and even their names, were designed to tighten and protect their unity. No one among the Ju/'hoansi was more important than anyone else. With no chiefs or headmen, key decisions were made by consensus, and a woman's opinion was as important as a man's. No one wanted to be seen as the best hunter, or the best anything. To do so could make others jealous, and that could undermine the unity of the entire group.

In the Nyae Nyae area—the six thousand square miles where we spent most of our time—were fifteen waterholes, seven of them permanent (although one was polluted by hyenas who, while trying to drink, had fallen in and drowned), and eight that might go dry if the rain was insufficient or the dry season came too soon. The area had no surface water, so except for hollow trees with leftover rain water, the liquids in the guts of game antelopes, and certain watery melons, these waterholes were the only source of water.

Near most of them was a Ju/'hoan encampment, but almost everyone who lived in any encampment had relatives or in-laws in others. Their groups visited one another, often bringing news and gifts. It took about a year for a gift or important information to cross six thousand square miles. But cross it did, because the people liked to keep in touch with one another.

It was the Old Way. They depended on one another. If a waterhole went dry, the people who lived there could be welcomed somewhere else. Or if two people weren't getting along, one of them could move somewhere else. About a thousand people lived in this area, and with very few exceptions, all were connected, not only to those in their own encampment, but also to others where, if in need, they could find support. It may have been the best example of unity—and therefore personal safety—that our species has ever achieved. No wonder the Ju/'hoansi were non-violent. No wonder they called themselves Ju/'hoansi. Unity was the most important part of Gaia's second rule.

But there were other parts to that rule. If you live on the savannah and sleep on the ground, you need protection from predators—which in the case of the people we knew were mainly leopards and hyenas. The Ju/'hoansi made shelters to sleep in, and these, as far as I'm concerned, said much about antiquity. No one else seems interested here, and maybe I'm wrong, but the shelters reminded me of nests made by our relatives, the other great apes.

All great apes except adult gorillas make nests in the trees. Here it might be interesting to note that the only tools attributed to apes are the sticks they push into termite mounds. But a nest is a tool, and much more complicated. An ape must weave branches together in a way that they will stay together as a somewhat cup-shaped structure, and then stuff it with leaves.

The people did this too, although they were now on the ground. A Ju/'hoan woman would break branches from a bush, stick them in the ground, weave them together in a dome-shaped structure, and stuff it with grass. If you cup your hand and hold it palm up, it looks like a great ape's nest. If you turn your hand sideways, it looks like a Ju/'hoan shelter. The image isn't exact by any means, but it's suggestive. The two kinds of structures are roughly the same and serve the same purpose.

The Ju/'hoansi made new shelters when they moved, and did so again if they returned to where they'd been before. They didn't reoccupy the old ones. Great apes do the same, although other kinds of animals use nests or shelters for as long as they can, even passing them down through generations.

A nest in a tree keeps you safe from predators, and a shelter on the ground protects you from behind where a predator would attack you. And if your shelters are close together, facing in all directions as did those of the Ju/'hoansi, someone will see the predator no matter how he approaches.

A shelter is quick to make and easy, too. If pests of any kind were in the old shelter, they won't be in the new one. And the necessary materials are right where you're standing. Who could ask for more?

Rule Two has a clause about permanence. If a certain practice works, keep it if you can. As far as shelters were concerned, no improvement was needed, so our lineage could have kept Rule Two the same way from the time our hominin ancestors came down from the trees all the way to the San on the Kalahari savannah. That would be forty million years.

A serious threat to our species would have been predation by lions, who had been hunting members of our lineage since both of us evolved. In many places, lions still hunt humans. The Ju/'hoansi, however, had made an agreement with the lions, which is not necessarily an anthropomorphic statement—attributing a human quality to lions—because it might be a leomorphic statement—attributing a lion quality to humans. The Ju/'hoansi, of course, never hunted lions, nor did they hunt any other predators. They had no reason to do so. But the lions had every reason to hunt the humans, yet they didn't. When I first saw this I was in my late teens and early twenties, so I thought this was normal. Only until I was older and wiser did I grasp what I'd been looking at.

And it's hard to believe. But my brother and a colleague, Claire Ritchie, made a study of the causes of death among the San. Going back about a hundred years, and including about a hundred people, they found accounts of only two Ju/'hoansi who had been killed by lions. One was a man who lived at the edge of the pre-contact area near a Tswana cattle post where lions killed cattle and conflicted with people, thus didn't reflect the cultures—both human and lion—of the interior. The other victim was a partly paralyzed girl who dragged herself by moving her buttocks as she once had moved her feet, and may not have seemed human to the lion who took her life. Otherwise, although people and lions encountered each other often enough, the encounters were calm on both sides.

Considering how lions behaved in other areas, the truce seemed astonishing. Dozens of times, lions came to look at us in the encampment, but only to look at us or sometimes to roar a warning, as if they wanted us to leave. But they let it go at that. Lions even looked down on our faces one night when we were so tired from traveling that we didn't bother to make a camp and slept in the open on the ground. We knew what they'd done when we found their tracks in the morning. The incident I mentioned earlier, where four Ju/'hoan hunters firmly but politely told a large pride of lions to leave their dying antelope, is a good example of how the Ju/'hoansi and the lions typically interacted. I know of only one theory other than my own to explain the truce, from a man who told me that if people didn't hunt the lions, the lions wouldn't hunt them. To him it was a no-brainer.

But here's my theory. Major evolutionary and environmental events often have to do with water, as did this truce between lions and the San. During the dry season there were no lakes or streams, no surface water. The only water to be found was in waterholes, and those with drinkable water were far apart. Thus, by necessity, waterholes served both people and lions.

The San and the lions—the two apex predators—hunted the same game in much the same way—but with one important exception: the men hunted in daylight, and the lions hunted in the dark, stalking their quarry for about the same distance before shooting an arrow if they were men or charging if they were lions.

Otherwise the groups were quite similar. They both consisted largely of relatives—the lions as sisters and mothers, daughters and aunts, while the people were connected through blood and marriage—and both had encampments in which the young were protected by some of the adults (always in the case of people, sometimes in the case of lions) while the other adults were foraging.

Both groups used the same area, several miles in all directions from the waterhole. People used it by day, while the lions slept in a pile in the shade of a tree—if a patch of shade is already full of lions, the newcomers lie down on top of the others. And the lions used it by night, while the people stayed close to their fires, leaving the encampment after dark only in extreme situations.

Aside from being different species, the two groups were almost alike, one for the day and one for the night. No wonder they understood each other and wanted their arrangement to stay.

If one group was endangering the other, the two groups would tangle until one would be driven away. The people could live with relatives at different waterholes, but not too many in any one place because this would challenge the local food supply. It's better to live at home than as a guest, and given regular conditions, residence elsewhere was not a first choice.

The lions had more to lose if friction developed. If they were forced to leave a waterhole, they'd have two choices. They would still need water, so they could either wander in the bush, perhaps alone or with a few others, looking for watery melons or hollow trees with leftover rain, or they could go together to another waterhole and try to take it from the resident lions. Of course, a battle would result. The winners would stay, and the losers who survived would roam the savannah.

This was as undesirable for lions as it was for people. Both did what they could to avoid it.

This lion/San relationship may be vaguely known to science because I wrote an article for *The New Yorker* about it as well as two papers for academic journals (assuming anyone read them). But this part of the Old Way was never investigated by wildlife scientists, largely because none knew of it while it lasted, and now it's gone. That's a scientific tragedy because the astonishing behavior seems to have been agreed upon by both species for the same reasons. Not only the Ju/'hoansi but also the lions used agreement and cooperation, all in keeping with Gaia's second rule.

CHAPTER 26

Gaia's Rule Three

Reproduce

Almost all the Ju/'hoan adults we knew were married. We knew no women who had never married and only one man, who for some unknown reason had never been able to hunt successfully—a prerequisite for marriage. Children could be betrothed at an early age, usually for unity advantages, but if they didn't like each other after they grew up, they didn't have to marry, and if they married anyway, they could easily get divorced and marry someone else.

No formalities were attached to divorce. The man and woman just announced they were divorced and went their own ways. The children usually stayed with the mother, at least while they were little, but maintained a relationship with the father. A few men had two wives and we heard of a woman with two husbands, but most people were married only once and only to one person.

Adultery was rare. There was no need to be sneaky. A woman we knew once decided to join a man other than her husband and marched off in front of everyone to go with him to his encampment. Her husband was hunting at the time, and when he returned he was enraged. He was about to start after the departing couple to confront the other man, but some of the men in his encampment offered to follow the couple to see if they could persuade the woman to change her mind. The husband agreed, and they did. The wife came back with the helpful men and rejoined her husband, perhaps advising him to change some of his ways. Everyone was pleased.

Marriage had certain requirements. The Ju/'hoansi were so careful about incest that even third cousins couldn't marry. Thus they were more careful than the rest of us—Darwin married his third cousin—and thanks to their marriage rules, studies show the Ju/'hoansi to have less genetic similarity than the people of any other culture.

Before a man could marry he had to prove that he could hunt, which meant he would be in his late teens. His bride was usually much younger and almost always had been chosen for him. She and her new husband wouldn't have sex until she passed the menarche, which due to a low-fat diet and a strenuous life, wouldn't happen until she was in her late teens. An elaborate ceremony was held when she did.

But not much of a ceremony was held for the wedding, which adults did not attend. The only people present would be the groom who would be in his teens, the little bride who would be much younger, and some of the other children in the encampment, including toddlers. They would chat by the fire in front of a shelter built just for the occasion, the girls on the women's side of the fire and the boys on the men's side, and soon they'd go back to their parents' shelters. After that, although to us the ceremony seemed almost non-existent, the two were married, probably for life.

For the Ju/'hoansi, the important thing was that the two had entered the married state. It was almost as if they'd become a different kind of people. If they got divorced and married other people, there'd be no ceremony because the two had merely changed partners. They were already in the married state. Older people, especially women after menopause, were vaguely considered to have left the married state.

After the wife reached the menarche, she didn't get pregnant for a while. The Ju/'hoansi had no form of contraception, but women who work hard physically and eat no fat are not overly fertile. A Ju/'hoan woman might be in her mid-twenties before becoming pregnant.

If a woman gave birth in the encampment, she would expose men to the weakening effect of her female power. So when her labor began, she would leave the encampment quietly, telling no one and attracting no attention. Only if this was her first child would her mother go with her. Otherwise she'd go alone. While she delivered, she wouldn't scream or cry no matter how much pain she experienced or how long the process took. A scream would certainly attract the nearest predator—few things could be more tempting than a woman in labor.

Five or six years might pass before she got pregnant again. Most children nursed until they weaned themselves, which delayed a second pregnancy. Obviously this had prevailed since the earliest times; rarely was a baby born too soon after another.

But it happened. A Ju/'hoan woman produced only enough milk for one baby, again because of her low-calorie diet and extensive exercise. Since her milk was the only available nourishment for her baby, if she were to have a second baby while still nursing the first, she couldn't feed both and would have to kill the newborn. This could be avoided only if a lactating woman who had recently lost a baby would take the new infant, but at the very most, only fifteen or twenty women would be in any one encampment, and infant mortality was rare. We knew no one who had done this. Killing a newborn was just as hard on Ju/'hoan women as it was on anyone else, but there was no choice. If a woman tried to nurse two children, both would almost certainly die from malnutrition.

When the woman learned she was pregnant, she'd know what came next. She'd know, while nursing the infant she had, that the child in her body was doomed. And when alone in the bush giving birth to the baby, she would steel herself and do the right thing, and lose one child, not two.

Although the Ju/'hoansi managed Gaia's third rule quite well, having kept our lineage on the planet for at least a hundred thousand years, they had the lowest birth rate of any modern human population. In the area where we spent most of our time, the population was one person for every ten square miles. So if the Ju/'hoansi were any example—having lived the hunter-gatherer life with great success for longer than any of us can now imagine—it would seem that women were never meant to be as fertile as most of us are today, reaching the menarche early and getting pregnant soon if we don't use contraception. We now overpopulate the planet because Gaia didn't foresee our modern lifestyle. If we keep this up for the next few hundred years, let alone a hundred thousand years, we'll be so crowded we'll be standing on one another. This may lead us to extinction unless Gaia finds a way to slow us down.

CHAPTER 27

The Present

We came from the Old Way. Our ancestors were the San—we're here because of their knowledge and their skills. But the Old Way is gone for our species.

Many Ju/'hoan people we knew are no longer living, and those in the next generations no longer hunt or gather, except perhaps occasionally. Instead, they are trying to be what they want to be—rural Namibians, like everyone else in the region. Many of the families we knew now live in a town called Tsumkwe, once just a little spring beside a massive baobab tree, at least two hundred years old. I believe the place was named for the spring.

The first vehicle tracks ever made to that spring were made by me in an army truck. I happened to be pioneering that day and was told to travel south-east, so I did. When I saw the spring, I stopped and waited for my dad, who soon drove up behind me. He investigated the spring, the first non-San person ever to do so. Soon the two other trucks with the rest of our expedition appeared, and we all got out and drank some water.

Many years later I went there again and saw a white person's house near the baobab tree. I was thirsty so I went to the spring to drink, but a white man came out of the house and told me I couldn't have any water because the spring was his. Within a few minutes of our very first encounter with pre-contact Ju/'hoansi, they invited us to drink their water. I saw that times had changed.

The baobab died prematurely, I think because the white man had been watering his lawn and drained the spring. By then, Tsumkwe had streets, a gas station, a few stores, some government buildings, a clinic, a police station, a school, a church or two, several bars, and the houses of people who worked there as police or clergy or health-care providers or business owners or government employees, some but not all of whom were Ju/'hoansi.

Traditionally, a Ju/'hoan person had only one name and possibly a nickname, but by then the Ju/'hoansi had two names, a first name and a surname. Like us, the surname was the father's name, so that a man named Tsamko who was the son of a man named #Toma became known as Tsamko #Toma. (Here, the # mark is a click.) And traditionally, the Ju/'hoansi had worn leather clothes, but by then everyone wore western clothes except for some women who wore Herero* clothes adopted from the first Dutch settlers. Few if any grass shelters were made by then because most people in the town lived in houses. In the outlying areas, people lived in rondavels like those of the Bantu-speaking peoples. Some people around Tsumkwe lived in shacks, others lived in small, shed-like apartments provided by the government, and still others lived in cottages that they rented or owned.

Some people had regular jobs and western-style educations. Some people were farmers with livestock and gardens. But many people were poor and relied on intermittent government welfare.

Some people had cars, or cell phones, or laptops, or all three. I now communicate by email with Leon #Oma Tsamkxao, the grandson

---

* The Hereros, a Bantu-speaking people, had encountered white settlers soon after the whites arrived.

of people we knew well. When I see an email from him, or when I click "send" on my answering email, I remember his grandparents, our months of difficult travel to find them, and the thousands of square miles around them which at the time were "unexplored."

Many people now disapprove of the Ju/'hoansi for abandoning the Old Way, wanting them to be as they were. The disapprovers don't want to hunt or gather themselves, they just want others to do it so they can watch. Therefore, a tourist lodge has appeared, where Ju/'hoan men dress up in leather loincloths and take tourists—even female tourists—on make-believe hunts. The tourists return to the lodge feeling they've experienced the Old Way, and so they have, just a little, if they took part in tracking.

At the lodge, the Ju/'hoansi hold simulated dances in the daytime, although the real dances were held on nights of the full moon. The tourists enjoy those too. Film-makers see the lodge as a place to mix incidental characters with background scenery, and some silly films such as "The Gods Must Be Crazy" result. My brother called the tourist lodge "a hunter-gatherer museum." He knew what he was talking about because, except for the time before Namibian independence when he was barred from the country for apartheid denial and interracial fraternizing, he spent most of his life filming and helping the Ju/'hoansi. He wanted for them what they wanted for themselves, to live like everyone else.

As for me, in my humble opinion, even if the lodge is a hunter-gatherer museum, it provides jobs, which is good, and if some of the old traditions are still visible, even if somewhat inaccurate and only for tourists, is that so bad? In the last fifty years we've changed too, and we non-San people sometimes resurrect our past, so why not resurrect someone else's past? But my brother thought that a hunter-gatherer museum misrepresented the San themselves and the new lives they had chosen, and of course, this was true.

The transition wasn't easy, especially at the start. But help did arrive—at first mostly from my brother, and now from non-profit organizations such as the Kalahari People's Fund, which is run by the anthropologist Megan Biesele who speaks fluent Ju/'hoan and is an expert on past and present Ju/'hoan culture.

Immigrants from China have recently discovered Namibia. Like the European colonials in India, Africa, and the Americas, and most especially like the German, British, Dutch, and Portuguese whites who once ruled southern Africa, Chinese people have arrived, literally by the thousands, to make their fortunes by exploiting its resources. Not only is the Old Way gone, the place where it began is being stripped of everything that reminded us of it.*

In part, I'm telling this to show what's been lost with the Old Way of the Ju/'hoansi—immunity from foreign exploitation, a stable population, a favorable view of non-violence, near-endless information about the natural world, observational skills now hard to imagine, and no pollution. But as far as the Ju/'hoansi are concerned, I don't mean this as a mournful statement. At the time of this writing they're not yet deeply affected by the Chinese invasion, and they no longer want their former life. Would you?

Could a little grass shelter take the place of your current residence? Could you walk barefoot fifteen hundred miles a year—that's roughly from New York to Austin—carrying a load that weighs as much as you do after you get to Memphis?

Would you want to give birth alone in the bush, perhaps with a leopard nearby? And would you be willing to kill the baby you delivered if you had to? With your low-calorie diet and your fifteen hundred miles a year, there'd be little chance of your needing to do this, but if you had to, could you?

---

* An open letter to the Chinese ambassador to Namibia from Dr. Chris Brown, CEO of the Namibian Chamber of the Environment, addresses the appalling number of wildlife and environmental crimes now in progress. "As Chinese nationals in significant numbers moved into all regions of Namibia," wrote Dr. Brown, "setting up businesses, networks, acquiring mineral prospecting licenses and offering payment for wildlife products, the incidents of poaching, illegal wildlife capture, collection, killing and export have increased exponentially."

Could you track a wounded antelope for many days, spending the nights beside a little fire, unarmed except for your spear, which is shorter than a leopard's reach? Forget the poison arrows—the poison works too slowly. Could you then walk all the way home carrying your own weight in meat? You'd have been hunting with others, of course, and you'd have discarded the hooves, some of the guts, and maybe the head, but even so, an eland weighs a thousand pounds, so if an eland was your victim, each of you might have to carry more than your own weight in meat. Could you do that?

And after doing all that work for all those days, taking all those risks, maybe finding no food and just a few sips of water although the temperature was 120°F in the shade, would you be willing to accept a few handfuls of meat as your gift for the eland you've just provided, rather than owning the carcass?

I'd say the answer is no. We wouldn't want that life even though it kept us on the planet for thousands of years. But compared to modern lifestyles it was tough. If the Ju/'hoansi would rather live in rondavels or houses, have jobs, have farms, make money, buy food, and sit around in bars like other Namibians, it's not hard to understand.

CHAPTER 28

The Future

We're a tag-end species on this planet. Life appeared 3.8 billion years ago. Compared to that, we appeared a few seconds ago, and we won't be here for long. We won't be like the big dinosaurs, who were here for 145 million years, let alone the waterbears, still here after 530 million years, because even if we're still present in five million years, the sun will burn out. My astronomer cousin, Tom Bryant, offers the theory he prefers for how this will happen: "The sun will become a white dwarf," he says, "and our dry and frozen earth will slowly orbit what's left of its core, much as it always has."

Of course, our present population won't be here when that happens. "Here's a phone," we might say. "Call someone who cares." But maybe we do care. We were given a gift by Gaia, the gift of the natural world. We've removed ourselves from most of it, but whether we care about it or not, we own it, and if it fails before the sunlight stops we'll have ourselves to blame. Life is a delicate phenomenon. We should keep it if we can. But if we can't, that may not matter in the long run, and not because we don't care.

The universe is said to have upward of a hundred billion galaxies. Another estimate says five hundred billion. Each galaxy is said to have perhaps two hundred billion stars, some with planets. To support life as we know it, a planet must have certain elements, such as nitrogen and carbon, and it must have water as a liquid, probably not as ice or steam (although waterbears could live with ice or steam), so the planet must not be too near or too far from its star. Not all these requirements are met by every planet, but with one hundred billion galaxies to choose from, each with two hundred billion stars, and if the laws of probability mean anything at all, life might be found on at least one hundred billion planets. Even if only 1 percent of those planets hosts life, that's a billion other planets that could have developed into something like ours.

It's not possible that we're alone in the universe. Far from it. And one can imagine that on some of those planets, the apex life-forms have not lost touch with the others around them. Would that we could do the same. We won't or we can't, but we could send a message. If we send that message to a planet where some fish are trying to live on the land, and if the speed of light is as fast as we think, they'll receive our message when their apex candidates are on their hind legs, wondering what happened to their body hair and if they did right by leaving the trees.

Do as we say, not as we do, we'll tell them.

Coda

When one thinks of the amount of information we now have about the life-forms of the earth, let alone the amount we now know about the universe, one should also think of the meticulous work that has gone into gathering this information, or in other words, the careful work of scientists. In a way, they resemble our hunter-gatherer ancestors in that they depend on accuracy and do whatever necessary to achieve it. So remembering the millions of life-forms now on earth, to say nothing of the endless chain of ancestors that gave us the present life-forms, virtually all of which were discovered and described by scientists, it helps to consider an example of scientific study.

To provide an example, I wanted to find a scientific paper completely at random, just one among dozens of equally important papers, so I groped in the dark on a shelf that holds a stack of journals and took just one. When I turned on the light I was holding *Journal of Mammalogy*, Volume 95, Number 4, August 2014. It fell open at page 774, the third page of a paper about foxes.

The paper is by Barbara L. Langille, Kimberly E. O'Leary, Hugh G. Whitney, and H. Dawn Marshall, and the title is "Mitochondrial

DNA Diversity and Phytogeography of Insular Newfoundland Red Foxes (*Vulpes vulpes deletrix*)."

One can only imagine the hours, days, and months that went into gathering the information, the outcome of which, as expressed in the paper, is that a baseline was established "for continued investigation of population demography, genetic structure, and adaptive genetic diversity in island Newfoundland red foxes, a population of interest from both ecological and wildlife disease perspectives."

The study explored "genetic variety at the mitochondrial control region in 189 foxes," finding support for an earlier hypothesis that "indigenous red foxes in North America are derived from disparate refugia isolated during the Wisconsin glaciation." Newfoundland, according to the paper, was "recolonized (by red foxes) via a northern glacial refugium via Quebec or Labrador rather than via an Atlantic or southern route."

The fox population was also found to be "the only endemic focus in North America of the endoparasitic nematode, French heartworm (*Angiostrongulys vasorum*)."

"So what?" you may ask. Yet it's from such intricate bits of hard-won, carefully collected, extensive evidence that we understand our world. Thanks to the study of Newfoundland foxes, we may better understand the movement of post-glacial animal populations, and also that French heartworm may be on its way to the United States. We laymen may not care that an animal population emerged from northern refugia rather than by southern routes—although that's the key to possibilities that wildlife scientists will certainly value—but we'll care if our dog gets a new kind of heartworm that her present medication won't prevent. It's something to consider that everything we know about the world and all that's in it, from the formation of the Himalayas to the function of the fibers around the nostrils of a platypus, was discovered bit by bit with this kind of concentration and care.

Acknowledgments

My deepest thanks must go to the Ju/'hoansi for their friendship, their generosity, and their understanding. My gratitude is undying.

For facts and fact-checking about the universe, I thank my astronomer cousin, Tom Bryant. For reading this manuscript, finding inaccuracies, and offering important suggestions, I thank Gary J. Galbreath, Marcus Baynes-Rock, and Mark Witton. These gentlemen, all scientists, could not have been more helpful, so if my book has mistakes, as it very well might, I added them later. For expert editing and unlimited patience as we prepared this manuscript, I thank Kendra Boileau, not only for her insight and her editorial skills, but also for her friendship and hospitality.

For review of this book, for many suggestions, and for invaluable support, I thank my dear friend Sy Montgomery. For thirty years of help and support that allowed me to write without interruption, I thank my dear friend Anna Martin.

For information about contemporary San and for valuable advice, I thank my dear friend Megan Biesele. And I offer special thanks to Vickii Lovelady. After hearing a reading of part of this book, Vickii said it should be used in science class because it isn't boring. At the time, she was in seventh grade, so I found her statement encouraging.

For their support and understanding, I thank in alphabetical order Bob Kafka, Saibhung Khaur Khalsa, Saibhung Singh Khalsa, and Stephanie Thomas. I thank Steve, my late beloved husband. He read the book too. He was a scholarly, highly educated person, and rather than exclaiming with joy at every well-written sentence, which would have been nice, he frowned with concentration, wondering if he could trust the information. That was nice too. If he didn't trust the information, I could try to fix the problem.

I thank my little dogs, Kafka and Čapek, who sleep close beside me and keep me warm at night. I thank my gray cat who purrs when he's dreaming. I thank my other gray cat who almost every night provides me with headless carcasses of mice, or at least with some of their internal organs. I thank Claude, my husband's cat, who stayed beside Steve, his hand on Steve's, during Steve's illness.

I thank the deer, the bears, the coyotes, the foxes, a bobcat who lives in our woods, the wild turkeys, the hummingbirds who visit my feeders, also the Kalahari lions who visited our camp, the hyena who came part way into my tent but didn't bite me, the leopard in northern Uganda who, although it was night and he may have been hunting, let me walk by him in safety. I thank the wolves I knew on Baffin Island, the oaks by my house who showed me about withholding acorns, the vine who showed me how a vine works, and many other non-humans too numerous to name who showed me much about the natural world, who enriched my life, and will live in my heart forever.

END